CITIES AND ARCHITECTURE UNDER DEBATE
Europan

Edited by
Chris Younès et Alain Maugard

Parenthèses

Acknowledgements

This book is the result of a collective effort by all those who have helped to build Europan over the last thirty years; thank you to the 759 European cities who put forward a site for the competition, to the juries and experts from the 14 sessions, to the 24,391 European teams that took part and to the 2,108 teams selected by Europan.

This book would not have been possible without the dedication of Isabelle Moulin and Louis Vitalis.

www.editionsparentheses.com
ISBN 978-2-86364-965-7

PREFACE

EUROPAN: GENIUS LOCI

Hélène Peskine

Puca Permanent Secretary

For thirty years, this competition has been providing young professionals in the fields of architecture, urbanism and landscape architecture with an opportunity to showcase their creativity and to take part in a great human adventure. This adventure will undoubtedly have a lasting impact on their journey as a designer, might set them on the path towards a theoretical approach inspired by Europan. This program, which brings together several European countries, treats these young designers as globetrotting citizen-practitioners tied to a shared culture – the culture of the Roman town – and an urbanism that is primarily managed by social and environmental public authorities.

The Europan competition is a unique opportunity for dialogue between young architects, urban planners, landscape architects and urban developers. Over a period of two years, the competition places the project and its gradual implementation at the heart of a generous and formative process. This process allows for an unprecedented level of dialogue between designers and clients, in ten to fifteen countries, about dozens of prospective sites that will be resolved by means of a collaborative project.

The project, rooted in its site and managed in conjunction with the local community and the territory and its stakeholders, is the cornerstone of this dialogue. It is the object that brings together these disparate forces so that they can progress as one.

In 1972, Puca, the French government body responsible for carrying out research in the fields of architecture, construction and urbanism, created the Programme architecture nouvelle (New program for architecture), or Pan. In 1988, Pan became Europan, drawing on the wealth of practices, sites and cultures of its new member countries. These days, we talk about an "augmented" Pan. This neologism, borrowed from the field of 3D printing, recognizes the structural impact that digitization has had over the last thirty years on the design and guidance of territorial projects. Digital technology has already had a huge impact and, together with the relationship to materials and the territory, and the challenges of climate change and social solidarity, is set to remain at the heart of the questions and responses of future Europan sessions, as they try to predict the major changes in our professional practice.

Puca's raison d'être is to design and carry out action research and experimentation programs that unite theory and practice and prioritize dialogue between stakeholders – from the designer, to the client or contractor, to the end user. The ability of Europan to keep pace with, and often anticipate, changes in architectural and urban thought and practice over the last thirty years is something we can be proud of.

This book was conceived by Chris Younès and Alain Maugard as both a prospective and a retrospective work, highlighting the work that Puca continues to do, after nearly fifty years, in carrying out assessments, archiving, and research for the benefit of so many people. So, what can we learn from the Europan generation? It was among the first to predict several transformations in how we build and design the city and the territory. Originally focused on housing, it pioneered the modern urban project, and guided designers and decision-makers towards new types of collaborative project, most notably promoting the possibility of projects that evolve over time and in space, the "landscape project" and the participative project. We believe that, for many years to come, Europan will continue to be thirty years ahead of its time!

INTRODUCTION

A EUROPEAN FAB LAB FOR THE TERRITORIES

Alain Maugard and Chris Younès

Europan's previous incarnation was Pan (Programme Architecture Nouvelle), which was established in 1972 as a reaction to the architectural policy of standardised production and the dominating influence of the Grands Prix de Rome on architecture. Its aim was to disrupt the architectural power structure by paving the way for a new generation (which is why the competition was reserved for entrants under the age of forty), whom Pan called on to "build the revolution".

Following the success of its first ten years, Pan asked its European neighbours to join the program, so that they could share in this revolutionary ambition, pool their sites and "provoke" teams of young European architects. This was the start of Europan, a genuine melting pot of a shared culture.

The central thesis of this book is that, over the course of its thirty years and fourteen sessions, Europan has anticipated transformations, imagined futures, and invented new approaches – essentially it has always looked ahead, and with great acuity. We are so convinced of its perspicacity that we nearly called this book *Europan, 30 Years Ahead of its Time*. If we want to be less categorical about it, what we can say with confidence is that Europan has always opened up questions about the city and architecture to debate, and that it has always remained focused on a vision for the future. That is why the future is the central focus of this book: the future of cities, the future of action, the future of our professions.

However, in order to tackle the future, we first need to learn from Europan's fourteen previous sessions, to understand how they relate to current debates and how they can be used as the basis for prospective analysis.

The first two parts of the book, written together with Céline Bodart, are a scientific analysis of the theoretical and practical output of Europan. Here, we draw out the central elements of the transformations that have taken place and revisited a few standout projects from the many award-winning entries. These two parts also investigate the methods used to devise the competition briefs and commissions and the responses provided by the architects, urbanists and landscape architects, as part of a co-constructive approach. There is no doubt that Europan has produced a way of working for clients and project managers, enhanced by the participation of residents and citizens, that is successful because it draws on the efficiency of a collaborative approach that will only increase.

For part three, we assembled a group comprising Didier Rebois, Bernard Reichen and Xavier Bonnaud, each of whom has played a role in the history of Europan, to discuss the developments that will shape the future. Energized by this discussion, and with the help of Sylvie Groueff, we explore the trends of tomorrow outlined in this *Europan vision* for the future; a vision that is constantly being reinvented (as per Bernard Reichen's "eternal youth") because, with each new session, a new page opens on the future, leaving the previous pages intact.

Over the years, Europan has become a unique laboratory for experimentation, following and leading the metamorphoses of European cities.

EUROPAN TOPICS

- **E15** – Productive Cities 2 : Resources, Mobility, Equity
  2019/2020
- **E14** – Productive Cities
  2017/2018
- **E13** – The Adaptable City 2
  2015/2016
- **E12** – The Adaptable City
  2012/2014
- **E11** – Resonance Between Territories And Ways Of Life
  2010/2012
- **E10** – European Urbanity : Inventing Urbanity
  2008/2010
- **E9** - European Urbanity: Sustainable city and New public spaces
  2006/2008
- **E8** - European Urbanity : Strategic projects
  2004/2006
- **E7** - Sub-Urban Challenge
  2003/2004
- **E6** - In Between Cities
  2000/2002
- **E5** - New Housing Landscape
  1997/2000
- **E4** - Constructing The Town Upon The Town
  1994/1996
- **E3** - At Home In The City
  1992/1994
- **E2** - Living In The City
  1990/1991
- **E1** - Changing Lifestyles And Housing Architecture
  1988/1990

PART ONE

NEW POSSIBLES TO EXPLORE AND EXPERIMENT

Chris Younès and Céline Bodart

Architectures, cities: the plural form of the key concepts we aim to discuss here could certainly be considered Europan's initial motivation and recurring theme. Because the urban transformations that are currently underway are simultaneously societal, environmental, political and economic, they require us to engage *multiple* ways of seeing, perceiving, imagining and designing the potential future of European cities and territories. We need *multiple* architectures to question what our urban milieus have been and could still be. As an ideas competition with implementation, Europan has worked for thirty years to explore plural forms of urban imagination, while at the same time working on the ground with practices and in the discourses. It is around this theme-site-project triptych that the Europan competition has built and is still (re-)building its own unique brand: every two years, as part of a *themed* session, young architects, urbanists, landscape architects and others are invited to think in terms of *projects* on the urban future of *sites* from across Europe. That is what makes its trademark; always the same pattern but revisited and readjusted with every session. But while the pattern is repeated, what it can produce for each session differs. It is from the close interdependence and strong synergy between these three elements (themes, sites and projects) that the competition's structural pattern has produced Europan's plural dimension.

A PALIMPSEST OF SESSION THEMES AND WINNING PROJECTS

To reflect on the history of this very unique competition is to appreciate how this plurality has become part of a much needed continuity, or, more specifically, a continuous evolution. The diversity of the themes proposed by Europan has resulted in a kind of palimpsest of reflections on the future of urban and architectural practices. From small-scale housing to territorial projects, from questioning ways of life to rethinking the living, the conceptual shifts set in motion by each session are felt in the sessions that follow. This active and effective memory of the reflections fuelled by Europan finds its clearest expression in the evolution of its themes – a stratified evolution, but one in which the strata interact with one another.

Each competition session starts with the selection of a theme, a catalyst for invention that reflects contemporary societal and architectural shifts while also pursuing the lines of thought that were examined and explored by the previous sessions. It is these shifts in the stratification of ideas and collective inventions that characterize the history of Europan. So dense and dynamic is this history that we have decided to explain it in the form of *strata*, attempting to capture, explore and convey that which is specific to each stratum, as well as that which connects it to the rest. Each of these thematic strata should be seen as an articulation of the issues facing architecture, the city, urban areas and their multiple territories; each stratum attempts to express, and make shareable, the impetus behind the need to rethink how we inhabit contemporary urban milieus, their possible meanings and forms, and their modes of fabrication and production.

This stratified history is told via the construction of four thematic strata: 1) the question of housing and its plural architectures; 2) a definition of urbanity that can be tested against and enriched by its experimentations and multiple situations; 3) a renewed attention to the issues of the milieus and the living, unfolding multiple synergies between nature and culture; and 4) a socio-political reengagement with design practices. We developed each of these strata following an in-depth examination of the discourse produced by and about Europan, highlighting the multiple voices within it, whether these belong to internal representatives or external experts.

STRATUM 1: **LAUNCH OF EUROPAN, TRANSFORMATIONS AND PLURAL ARCHITECTURES**

Following on from where Pan (Programme d'Architecture Nouvelle, 1972-1987) left off, Europan launched its first Europe-wide competition of ideas in January 1988, asserting, from the outset, its desire to be a catalyst for architectural innovation. Although the theme for this inaugural session "Changing Lifestyles and Housing Architecture", explicitly continued the Pan program's focus on residential architecture, the description of its challenges also sought to differentiate Europan somewhat. The theme of housing was featured as an issue to be thought afresh: "housing [...] is, to some extent, blocked, stuck. It no longer provides the range of services that its users have a right to expect. Our ways of life, technology and societal structure all seem to foretell that our culture of inhabiting is undergoing a complete transformation."

The theme of Europan 1 drew attention to the fact that the architectural situation is fixed, blocked, and therefore problematic, and the whole point is that this situation should be addressed directly to those who are capable of responding to it. In other words, even at this early stage, Europan was asking architects to pay attention to the profound changes that were materializing in various aspects of the built environment and, at the same time, determining the future of their practice. The theme of Europan 1 marked the beginnings of an ethical and political commitment that would grow stronger and stronger as the competition progressed.

The speculative openness of Europan was also clear from the outset. In placing the emphasis on what current developments foretell, this theme of changing lifestyles set out the transformations that were underway but also, most importantly, those that were yet to come. Uncertainty about the future was framed as an input into the architectural project: a call for seeing and shaping new possibles, within and beyond the hesitations of a world still in-the-making. Of course, this stance resonates with the various experiments with more or less utopian design that characterized architectural and urban thought in the previous century. Europan's project inherited this desire to disrupt how we used to think and, from the end of the eighties, invited new generations of architects to do the same: to anticipate, invent and respond to the transformations that were beginning to take place within European architectural culture. Ambitious from the outset, Europan had one central belief: *multiple* architectures are needed (a need expressed as far back as this first theme) to reorient design practices toward the invention of new forms of inhabiting.

STRATUM 2: **URBANITY AND THE PROJECT AS ASSEMBLAGE**

Calling for an end to planning practices in which economic opportunism too often took precedence over "real, long-term urban policies", the second Europan session ("Living in the city, re-interpretation of urban sites") reminded participants that "housing still makes up most of the urban landscape". The aim of the competition was, at that time, to pose "fundamental questions about housing and the city and their relationship to one another"; it saw the need to "study housing in its environment, to consider its links with other urban components"[1]. The housing issue shifted towards, or rather was interlinked with that of urban forms of inhabiting. Here, the notion of *interlinking* meant rethinking housing "within a process of urban recomposition", in terms of its ability "to incorporate an environmental and democratic dimension" and to be "integrated into its environment"[2]. Europan, in using the

1 Direct quotes from remarks by Philippe Dehan, architect and expert for the Europan 2 session – available in the online archives on the Europan-France website: www.europanfrance.org.

2 *Ibid.*

concept of urbanity as a way to question how to bring together the many scales of inhabiting, was participating in a genuine expansion of the field of architectural invention.

Beginning with Europan 2, the question of urbanity begins to form the competition's basis for reflection and experimentation – albeit one that remains multifaceted.

Urbanity is an ambiguous and polysemous term that defines ways of being with one another, ways of coexisting. In Choay and Merlin's French dictionary of urbanism and development, the entry under "urbanity" includes references to the entries for "civility" and "city". According to Richard Sennet, "city" and "civility" share a common etymological root: "Civility is treating others as though they were *strangers* and forging a social bond upon that social distance. The city is that human settlement in which strangers are most likely to meet."[3] He examines the disappearance of the public spaces inherited from the eighteenth and nineteenth centuries, and the fact that, increasingly, the social units and scales of reference for the individual are domestic life and private socializing. Sennett goes so far as to consider the possibility that the citizen is losing his quality as a "public man".

The quality of city life cannot be separated from the multiple ways in which the individual and the collective are interlinked. Overcrowding and segregation, which are inherent in all forms of occupations or affiliations, in addition to social diversity issues, underscore the importance of approaches that encourage cohabitation between individuals and social groups. The home has always lead a kind of centering force, the cultivation of a small gap in a world of flows and networks. It is the fixed point from

3 Richard Sennett, *La Tyrannie de l'intimité* [*The Fall of Public Man*, 1976], Paris, Seuil, 1995.

which we move, and the place to which we return. That is why the home and its surroundings and links is an essential component of inhabited milieus. The establishment of a continuum between the domestic space and the territory acts as an antidote to the threats of segregated communities and isolation. The fixed points of spatial installations, whether permanent or temporary, whether chosen or endured, only function thanks to the lines of physical and mental mobilities with which they intersect and of which they form part. This intrinsic capacity to articulate the near and the distant operates by means of about-turns: towards the landscape and the public space as shared resources, towards a re-enhancement, against the background of environmental concerns, of non-motorized mobilities, such as walking or cycling, and public transport – not forgetting access to faster forms of transit and the little crevices that leave space, in this highly technical system, for stillness, for the unpredictable, for change.

In his essay "Transitions", Yorgos Simeoforidis (architect, critic, consulted as an expert by Europan) analyses the projects submitted to the particularly fecund third Europan session, which marked the point when representations of transition between public and private began to be seen as tools suitable for urbanity. In a sense, this essay describes not just session 3, but all the sessions that would follow, and follow on, from it, albeit with a greater focus on environmental responsibility. According to Yorgos Simeoforidis, "the theme of the third Europan session, "At home in the city – Urbanizing residential areas", presented young architects with the challenge of how to design public and open urban spaces, pushing them to respond to fundamental questions such as: "How should we think, today, about relationships between the private and public spaces in the city? What status do transitional spaces have, given the evolution of relationships between intimacy and urbanity?"

"Through their projects, architects are competing – more or less consciously – with the contemporary cultural origins of this question: the legacy of Modernism and the International Congresses of Modern Architecture (Ciam), the legacy of the city park, the legacy of the "landscape art" of the sixties and the legacy – if you can call it that – of the urban "strangeness" of recent decades. Each of the 85 projects [...] to a greater or lesser extent, comes under one of the four sub-categories of the Europan 3 theme, which is central in contemporary thought on the city, from the theory of the Ciam congresses, to the phenomenology of Team 10, to the urban anthropological perspective that underscores the importance of urban rituals and their spaces. These doctrines form the basis for a complex interplay in the modern metropolis between the use of urban space and the social organization of the city.

"This critical question can be used to interrogate the projects of Europan 3: to what extent do they take into account the cultural use of open and public spaces, the meaningful topology of the spatial articulation in relation to the community that inhabits it?"[4]

The legacy of these many reflections and experiments on the transitional spaces between public and private, the forms of articulation between intimate and shared, requires us to constantly update and relocate the challenges – to recall that which has already been done to actively and critically question that which is still left to do. Such a consideration echoes the argument put forward

4 Yorgos Simeoforidis, "Transitions", article published in the Europan 3 European results catalogue, Paris, 1994, p. 13.

by the philosopher Jacques Derrida: "This inheritance must be reaffirmed by transforming it as radically as will be necessary. [...] Inheritance is never a given, it is always a task. It remains before us."[5] So, although Europan invites architects to imagine new social and societal assemblages in terms of spaces, this invitation is not written on a blank page. On the contrary, Europan's decision to encourage persistence, to open up new perspectives, to re-engage some latent ways of thinking within urban and architectural culture, is what sets it apart. And it is in this same spirit that designers are invited to take part in it.

The legacy of this necessary continuum between the domestic space and the territory, therefore, encourages Europan to experiment. While the third session renewed and revisited some architectural and urban thoughts that considered transition spaces to be vectors of "plural citizenship", the theme of Europan 4 ("Constructing the town upon the town, transformation of contemporary urban sites") sought, more specifically, to place "upon the traces of the modern city" the forms of urbanism generated by it, and broke down the challenges into four key areas of investigation: "to situate, to transform, to inherit, to interlink". Similarly, Europan 5 and its "New housing landscape" and Europan 6 and its "In-between cities", revived the search for new representations, new figures of urban transition and architectural expression, moving the field of investigation to the vast peripheral areas fragmented by the proliferation of transit networks and their infrastructure, places where "a town shaped in hollow, composed of the gaps carved out by urbanization" has emerged; territories that "have long been socially and physically marginalized", but which should now be reconsidered

5 Jacques Derrida, *Spectres de Marx*, Paris, Galilée, 1993, p. 94.

as the strategic situations in which to invent new genres of urbanity. As these different urban configurations were explored through various projects, and as the sessions progressed, Europan built its own language to speak about these urban-architectural issues: diversifying and arranging the *rhythms* of mobilities; reorganizing the distribution of functions in terms of *proximity*; addressing the challenge of *shared spaces*; reconsidering the construction of architectural forms as a trigger *of urbanity*; revitalizing existing architectural and urban typologies, prioritizing *mixed*, *composite* and *hybrid* programs; making the choice of *urban complexity* and taking into account the rich but intricate multiplicity of public and private stakeholders involved. Listed in this way, the profusion of Europan terminology might seem overly theoretical, or even excessively prescriptive. However, it should be remembered that what makes Europan so unique is that it is constantly drawing on the diversity of its project contexts in order to write and rewrite its own structural pattern. It learns from what it produces. For example, when sessions 8 and 9 presented "European urbanity" as their research theme, they drew on the research carried out in previous sessions in order to put forward their own definition of the term "urbanity" as the basis for new reflections on the theme: "Urbanity is a political, cultural, social and professional concept that materializes in the form of spatial interventions at the interface between the urban project and the architectural project. It should be thought of as a shared way of experiencing the city and its functions, as a desire to reflect on the heterogeneity of its forms", but also as a "way of approaching urban space that encourages people to meet in the places they share: the public space."[6] From its diverse range of experimental project designs, Europan has crafted an *effective* definition of the concept of urbanity; a definition that relates to the effects of that which it has produced.

Constantly switching between the "theorization of practices" and the articulation of "theorized practices"[7], Europan's strategy is essentially based on the strategy of the *in situ*. And the evolution of the urbanity question as a thematic stratum of the competition is a clear indication of its importance. Through its project-focused research, it enriches the ongoing debate; through its themed discussions, it participates in the exploration of new possibilities, calling on designers to consider alternative relations between the local and the global, the shared and the individual, the human and the non-human. These assemblages require us to reconsider what is already there, in terms of both the traces of the past and its futures.

6 Extracts from the topic descriptions for Europan 8 "European urbanity and Strategic projects" and Europan 9 "European urbanity: Sustainable city and New public spaces".

7 With this dual reference we draw on and somewhat truncate the types of art theory proposed by Anne Cauquelin, a French philosopher, novelist, essayist and artist. Cf. Anne Cauquelin, *Les Théories de l'art* [1998], Paris, Presses universitaires de France, coll. "Que sais-je?", 2014.

STRATUM 3: **NATURO-CULTURAL SYNERGIES AND THE PROJECT AS ASSEMBLAGE**

The thematic stratum that has undoubtedly had the largest impact on the competition's evolution, both in terms of site choice and the successive iterations of the session themes, has resulted from the emergence of a new concept of nature.

Environmental concerns have highlighted the precarious nature of living environments and the finite aspect of planet Earth, pushing us to ask how human production can best accommodate tectonic, climatic and biological forces. Given the increased pace of change, the impact of our behaviours, and our common future, the way in which we approach and nurture our interdependencies is critical for the building and rebuilding of human settlements. In order to manage the underlying or overt processes involved, we need new forms of intelligence and sharing. This wake-up call has resulted in a genuine paradigm shift, placing the question of relationships, connections and other links between human and non-human environments at the heart of how we think about and design the contemporary world.

In the context of the Europan competition, this paradigm shift was first expressed explicitly in the fifth session, which invited architects and urbanists to imagine what the "new housing landscape" might look like. Links, relationships, porosities and other interactions between natural and inhabited milieus were presented as project contexts integral to the invention of new forms of quality urban space. In other words, the concept of nature was introduced, above all, simply as a tool for conciliation and reconciliation in areas that had been fragmented by previous waves of urbanization; a means, also, to mediate between the push for densification and the pull of urban vibrancy, between new kinds of shared spaces and inhabitants' overwhelming desire to have highly individualized homes. To see how this *other* concept of nature started to engage experimentation with new urban and architectural approaches, we need to take a look at the "In between cities" of Europan 6; a session that not only welcomed and pushed for the enlargement of the competition sites[8] but also encouraged local authorities to formulate their briefs *on the basis of* the environmental and landscape issues inherent in the sites they were submitting. For example, for the Sotteville-Lès-Rouen site (for which the study site measured nearly 25 ha), the city asked competitors to think about how to regenerate an area of land located next to a railway by making use of the characteristics of this infrastructural landscape: how to "showcase the daytime and nighttime spectacle of the marshalling yard, using an innovative landscaping approach and prioritizing simple solutions that respect the types of vegetation that grow naturally in this type of urban space."[9] Translating these

8 This was the first session to introduce, for each site, a "study site" (which measured up to as much as one hundred hectares in later sessions) and a "project site". This change to the initial format of the competition was made in order to better respond to the changes in cities' requirements, which were increasingly focused on issues that cut across planning strategy and the built environment. It reinforced and expanded the importance that the competition afforded to "the interweaving of scales".

9 Quote taken from the site presentation.

potential intersections between natural and cultural rhythms into physical spaces is exactly the objective that Europan 7 set out to achieve by inviting designers to rethink the forms and scales of urbanity in light of the digital and environmental transitions taking place in the contemporary world. Its "Sub-urban challenge" was a matter of assemblage between concepts of nature, environment and landscape; it asked participants to reconsider intervention processes in "taking into account the urban ecology as a whole" and to rethink project timescales as "long-term processes", spatio-temporal strategies capable of adapting to the multitude of stakeholders and interests involved. The following sessions aimed to explore these new avenues for urban and architectural reflection, even going so far as to make these environmental concerns a condition of positive urbanity. As the call for Europan 9 put it: "In order to design urbanity-focused projects that tackle the question of the role of the public space, we must re-situate this question within the context of sustainable urban development, that is to say, a spatial transformation that considers the environment and incorporates it 'sustainably' into the transformation process".[10] Once again, as part of Europan 11, "Resonance between territories and ways of life – What architectures for sustainable cities?" competitors were asked to imagine how the methods for designing urban-architectural spaces could become a sort of echo chamber for the environmental and geopolitical concerns that affect all European cities. The sustainable future of inhabited milieus was posed as a strategic challenge in which design practices must play an active role at all levels. Against a backdrop of global

10 Quote taken from the topic description for Europan 9 "European urbanity: Sustainable city and New public spaces".

societal change, Europan adjusted its objectives, while the projects submitted offered local and situated proposals of how to start addressing these new issues.

In response to the now recurring theme of possible intersections between the natural and the cultural, the projects experimented with links, resources and flows in order to establish the borders and crossings that could be used to shape a habitable world. In doing so, they also drew on ethical and political knowledge, purposes and positions. This naturo-cultural quest does not harken back to a nostalgic past, nor is it a rejection of the city; rather it is the search for alternative ways of making human settlement. What we need is a new politics of civilization, as Edgar Morin explains, in order to harmonize the rhythms of humanity with those of the universe. Nature, a constantly renewing force of genesis and metamorphosis, is the water, the earth, the air, fire, fauna, flora, the rhythm of the seasons, of day and night, of our heartbeats and our breathing, of birth and death. At once wild and domestic, it is both threatening and peaceful. Various forms of hybridization and symbiosis were explored in this interweaving of scales: eco-architecture, density carefully planned to preserve woodland and agricultural areas, planted streets, gardens and parks, microclimates, soil and biodiversity management, recycling... A whole new state of mind was emerging.

As this engagement was gradually being incorporated into the themes of consecutive sessions of the competition, it was also becoming a more prominent feature of the projects submitted. Starting in the early nineties, the projects "moved away from the use of nature as decoration towards an increasing appreciation of nature as a qualitative vector of a new kind of space and, finally, arrived at an understanding of nature as a dynamic force that defines the principle of its own future and its own ebb and flow, a principle that architecture can use to design and regenerate."[11] It is from the third session "At home in the city" onwards that architects really began to tackle these challenges in their projects. Although it still makes use of somewhat domesticated and mastered forms of "nature", the project *Entre pierre et jardin* [Weak Density] in Saintes[12] stands out thanks to its proposal to reinterpret the thin strips of the Bassompierre city block while sensitively reappropriating the site's history and layout. In order to insert new typologies that resonated with the unique texture of the project site, the project used a comb-like series of alleyways, protected by walls of reclaimed stone, to construct a new system of intermediate spaces that link the residential areas and reinforce the communal aspect of the whole block. This integration with the historical and material texture of the place is also achieved by a revival of the notion of the ground as something shared, linked, inherited; a dimension that the project team chose to reinforce by planting the common areas, creating green ribbons of private and/or shared gardens and allotments. Beyond the use of "nature" as a shared motif, *Weak Density* also paid particular attention to the environments bordering the project site, encouraging connection with the river via the construction of a public space designed to provide close contact with the water, its rhythms and its fluctuations (a floodable square, a floating bridge, etc.)

11 Frédéric Bonnet, Chris Younès, "La nature et le projet : enjeux et figures", in *Europan France 1988-2007 – Innover dialoguer réaliser*, Paris, Jean-Michel Place, 2007, p. 102-117, p. 103.

12 Project team composed of Armand Nouvet, Thibaud Babled, Laurent Berger, Marc Reynaud and Cyrille Véran (see p. 50).

As the sessions progressed, Didier Rebois (Secretary General Europan Europe) saw this kind of urban-architectural attitude give rise to a new breed of architect, which he called "sustainable landscape architect", by which he meant the "forceful, engaged, almost militant attitude around sustainable qualitative approaches to urban and architectural projects, [which] you find in some European teams, albeit expressed via different forms and ideas. This is not the slavish application of overly rigid principles of quantitative sustainable development. Rather, it is the challenge of finding how to transcribe, in the form of the urban space itself, the fundamental idea of a reconciliation between the city and the environment. The aim is to use an interactive approach, in which the incorporation of the laws of nature coexists with a vision for the urban project, to make use of the traces of existing forms in order to transform them, while also thinking about how the space will be appropriated by future generations."[13]

This possibility, introduced by the projects of the third Europan session[14], of alternative relationships with the concept of nature, was then interrogated and tested extensively and under various conditions by subsequent competitors. For some teams, this meant reinvesting in the topography or hydrography of the project site, interweaving its geographical contours with

13 Didier Rebois, "Questions / Hypothèses / Figures", in *Europan France 1988-2007 – Innover dialoguer réaliser*, *op. cit.*, p. 214-237, p. 226-228.

14 The French jury for this session included a landscape architect (Michel Corajoud) for the first time; this decision to expand the professions invited to discuss and evaluate the projects submitted for the different sites reflected the transformation that the competition organization was undergoing, incorporating concerns about the environment and landscape into its debates and reflections.

typo-morphological characteristics: for example in Chalon-sur-Saône for Europan 8, the project *Au fil de l'eau*[15] restructured the system of spaces (public, private and transitional) on the site by constructing a network of multi-use streets that connect to a new public space that runs along the river bordering the neighbourhood and its surrounding areas. For others, it meant designing new articulations between the built and non-built environment and an opportunity to re-interrogate the dominance of certain housing models that have a negative impact on territorial resources and landscape qualities, such as the intense development of suburban housing: in this vein, again for Europan 8 but in Hénin-Carvin this time, the project entitled *La Nature au quotidien*[16] envisaged an alternative to the housing estate by designing a hybrid structure that arranged residential units within a park composed of public gardens, defined as an open space for all. Later, when the theme of "The Adaptable City" was proposed for the second time for Europan 13, many projects took advantage of the landscape features of the sites – which were often very vast – to regenerate the urban qualities of the centres of small- and medium-size towns, enlarging their transformation processes to the scale of public space sequences. For Nathan Starkman, president of the French jury for this thirteenth session, one of the features of these project proposals was their attention to the sites' geo-morphological forces, and their use of architecture to transform the impact of key public spaces: these proposals "were multiple and broadly distributed, divisible, and adjustable over time. Landscape, sustainable development and long timeframes took centre

15 Project team composed of Gaëlle Cuisy, Karine Martin, with Anthony Brook (see p. 51).

16 Project by Fabien Gantois (see p. 52).

stage; the processes were in the background, embedded in the implementation of the proposals."[17] For example, in Saint-Brieuc, one of the two prize-winning projects, *Boulevard de la mer*[18], structured its intervention around two main paths through the landscape: the first crosses and reinforces the network of public spaces in the town centre, while the second transforms a former railway into a "seaside boulevard". This plan of intersecting axes was accompanied by a proposal for a "toolbox of contextualized tools for intervention". This toolbox encouraged reflection on and experimentation with the heritage value of these spaces, the construction methods that could be used on the site, and the potential modes of transport and approaches to property management.

Since it first appeared in the Europan discourse and its project entries in the early nineties, the concept of nature has evolved significantly, giving rise to a multitude of new project approaches and just as many representations of urban and architectural design. In an article entitled "La nature et le projet: enjeux et figures", Frédéric Bonnet and Chris Younès revisit the projects submitted in the first twenty years of the competition and focus on the emergence of five key trends: 1) the desire to *frame* nature by and for a specific architectural purpose; 2) the continuation of an antagonistic approach, a semantic and spatial *opposition* that reveals a certain "resistance to incorporate the central dynamics of nature, and the desire to leave the two imaginaries of urban architecture and nature clearly delineated"; 3) experimentation with new *alliances* "in all their forms: textures and fabrics, geographical

17 Nathan Starkman, "Des projets entre réalisme des contextes et ambition des thématiques", in *Europan 13: Catalogues des résultats*, 2016, p. 73-74.

18 Project team composed of Cyril Breton, Pierre-Olivier Carpentier and Maxime Genevrier (see p. 53).

foundations, natural tectonic, atmospheric and biological forces", the intellectual impact of which – often based on the power of obvious paradoxes ("to relate and differentiate, to delimit and interlink") – is hard to reconcile with its potential and necessary materiality; 4) the re-emergence of *references* to landscape-focused models of urbanism that, when combined with the actual project context sometimes hinder the social and societal principles on which they are based; 5) and finally, a representation fuelled by the image of "untamed nature as a resource and an environment for livings things", a vitalist force, flirting between an aesthetic of the sublime and naturalist metaphors, imagining the possible spatial junctions between controlling nature and setting it free, but, above all, a representation primarily used as a simple illustration: "the process of growth, the dynamics of transformation, and their various combinations"[19]. Each of these representations forms part of a field of possibility that is both strengthened and undermined by multiple (false-) imaginaries. In this sense, it is important not to take any representation or position for granted. In terms of architecture, the power of this concept is that it requires us to perpetually redefine it, paying close attention to the specific spatio-temporal context.

Accordingly, the fourteenth Europan session, "Productive Cities", continued to this ongoing experimentation by encouraging reflection on the "synergetic" representations of the concept of nature[20]; project

19 Frédéric Bonnet, Chris Younès, "La nature et le projet : enjeux et figures", *op. cit.*, p. 116-117.

20 In light of the results of Europan 14, Didier Rebois and Chris Younès identified the emergence of six "synergetic threads": synergies between nature and artifice; between local resources and sustainable urbanisms; between transport and production; between mixed

representations that investigated the possible and effective interlinkings between the parts and the whole. Constantly moving back and forth between scales – from the ecosystem to the neighbourhood –, these representations provide us with an opportunity to rethink the process of transformation of inhabited milieus by examining the multiplicity of its agents and components, because they are natural *and* artificial, social *and* political, human *and* non-human, *and... and...*, and more specifically their modes of relationship, collaboration, articulation, interaction. The synergetic aspects of the concept of nature lead to an understanding of the urban as "a collective metabolism"[21]. This new understanding calls for the establishment of new project communities and for experimentation with new *types* of assemblages between knowledge and know-how, between *make-with* and *make-together*.

uses and practices and urban architecture; between public and private production; and synergies between scales and rules in space and time. Cf. Didier Rebois, Chris Younès, "Villes productives et synergies", in Roberto D'Arienzo, Chris Younès (dir.), *Synergies urbaines : pour un métabolisme collectif des villes*, Genève, MétisPress, coll. "VueDensemble Essais", 2018, p. 207-220.

21 Taken from the title of the previously cited work by Roberto D'Arienzo and Chris Younès.

STRATUM 4: **FROM THE SEARCH FOR NEW URBAN "TYPES" TO NEW "WAYS" OF MAKING**

Another key characteristic of the calls for ideas launched by Europan is their aim to encourage competitors to imagine and design new "types". In a certain sense (and to intentionally labour the repetition), we could say that this competition interrogates the possibility of new *types* of European cities, encourages the invention of new *types* of urbanity, via experimentation with new *types* of mixed uses, functional programs, and relationships with environments, translating itself spatially by means of new *types* of connections, articulations and interactions, deployed via the design of new *types* of public spaces and, above all, shared spaces. This is an exaggeration, of course; and yet, in terms of the many texts published before and after each session[22], these *types* and the questions that they raise make up a significant part of the Europan discourse.

Since the start of the competition, these *types* have created a genuine productive indefinition. However, although this "need for a new type of urban planning,

22 We are referring here to the thematic statements of intent published at the start of each session, but also to the results catalogues and associated publications (special editions of the journals *Traits Urbains* and *Urbanisme*). This large corpus includes articles produced by the competition organizers (the Europan Europe Scientific Community and national secretariats) and articles written by experts that were specifically commissioned to evaluate the submitted projects and prepare the juries for each competition session.

a new type of architecture" persisted as the sessions progressed, these new *types* are now required to be even more "active" and "performative"; a performativity that Kristiaan Borret (architect, urbanist, member of the Europan Europe Scientific Committee) believes requires architects and urbanists to be capable of "carrying out initiatives in the life of the real city that are backed by civic engagement"[23]. Having expanded the scale of intervention (from the residential to the urban; from the urban to the territorial), Europan invited designers to reconsider the scope and the specificity of their own interventions.

At the start of the 2010s, in an economic and political climate undergoing complete transformation, Europan chose to tackle this new fragility of inhabited milieus by broadening the scope of reflection and experimentation to include the *multiple* urban practices that make cities *and* societies. As Frédéric Bonnet explains, "this required a new attitude from designers: they needed to develop a dialogue with residents and politicians, they needed to be prepared to come down to earth and get involved. They certainly needed to find a *way* of doing things that brings these protagonists together." This could be, according to him, "one way to address the current fragility of European urbanity", which also requires us to "consider the link with the economy, and not just land economy, but also local economies, which means working differently and thinking differently about the *way* we build the city."[24] In recent years, Europan has multiplied the

23 Kristiaan Borret, taken from his introduction to David Saxby, "L'architecte entrepreneurial", in the presentation brochure for the E13 topic "The Adaptable City 2", 2013, p. 8-10, p. 8.

24 Frédéric Bonnet, "La ville partagée : post-hédoniste et solidaire ?", in the presentation brochure for the E13 topic "The Adaptable City 2", 2013, p. 12-14. Our repeated emphasis on the word "way".

challenges involved in developing new *types* of urban spaces, by pushing participants to imagine and experiment with new *ways* of constructing them – new ways of producing the urban form, of creating environments and, also, new *ways of being* architects, urbanists and landscape architects. This recent update to the thematic challenges proposed by Europan invokes the ethical responsibilities of designers, while also inviting them to "re-appropriate the political question, to emphasize the political dimension of all decisions and the power of politics, whether in the sense of socio-economic power, the civic power that starts with on-the-ground initiatives, or politics itself, which is responsible for taking decisions that are in the public's interest."[25] The need for this kind of ethical and socio-political reengagement of urban and architectural practices was particularly evident in session 13 of Europan, "The Adaptable City 2". Faced with the increasing complexity of stakeholder networks, dwindling resources, limited funds, and the constant nagging worry about an environmental equilibrium under ever greater threat, we need to figure out how, in context, the combination of these initially paralyzing conditions can "inspire new project approaches". In a certain sense, the paradox of this challenge could be viewed as a way of "crafting hope at the edge of the abyss" to use the expression of the philosopher Isabelle Stengers[26], or "staying optimistic" as Frédéric Bonnet says. It is against this background that the idea of the project as a negotiated process is gaining ground in the Europan culture of thought and practice; it's another *way* of crafting urbanity, staggering

25 *Ibid.*

26 Cf. Isabelle Stengers, "Fabriquer de l'espoir au bord du gouffre : à propos de l'œuvre de Donna Haraway", in *La Revue internationale des livres et des idées*, n° 10, March 2009, p. 24-29.

its implementation rhythms over uncertain times, and considering not "just the intellectual and social skills of citizens, but also the technical and manufacturing skills of the building industry"[27]. This process and its negotiations will not simply materialize as some kinds of "miracle solutions"; they are ways of making with *and* making from a situation, from which new and unforeseen social, economic, political and environmental issues emerge. The aim of these project approaches is to "help us think about how to achieve a reconfiguration, how to draw on what we have in common, overcoming the very marked differences that have appeared in recent years with the emergence of neighbourhoods where people experience extreme economic, social, linguistic and cultural instability, have no right to speak, have no capacity to organize themselves or simply to mobilize around questions of development."[28]

These considerations came to the fore as part of the reflections and experiments carried out on the theme of adaptable and productive cities of the more recent Europan sessions, but they were not without precedent. Early on, the Europan competition encouraged participants to engage their projects with approaches to urban design that were *sensitive* to the multiple and changing times of its fabrication methods. Highlighting the way in which "the wide range of people involved, the mix of private and public stakeholders, the programmatic instability and the changing practices all affect the urban-architectural space", the fifth session, "New housing landscapes" was already encouraging competitors to design these new types of spaces as "a shifting process, with a constant

27 Frédéric Bonnet, "La ville partagée : post-hédoniste et solidaire ?", *op. cit.*

28 *Ibid.*

back and forth between the overarching ideas, multiple interpretations and staggered implementations." In other words, being able to "invent a project that is sufficiently robust that it can be implemented over a long period of time, while incorporating a dynamic of constant negotiation between the urban form, the program, and the architectural solution." These process-oriented and negotiated aspects of the project, presented as *ways* of "managing complexity", were taken up again in the sixth session, but as strategic areas of research via which to explore the multiple dynamics that make up contemporary urban environments; dynamics that are simultaneously social and spatial, particularly those between urban functions and services, which provide "the opportunity to articulate – in various combinations depending on the context – composite programs that mix housing, work, leisure, and transport". Again, for Europan 7, process and negotiation were presented as raw materials for the design of "new sustainable residential neighbourhoods". It was specified that competitors should "explain how their overall vision forms part of a spatio-temporal strategy that takes into account local construction methods"; to show how they intend "to establish spatial features to give coherence to the urban project while developing a series of programmatic interventions and appropriations together with their multiple stakeholders, as opportunities arise". In order to rethink architectural and urban interventions in terms of spatio-temporal strategy, Europan 10 continued in the vein of previous sessions, requiring participants to "Invent urbanity" on the basis of three kinds of intervention on, and with, the existing environment: "regeneration, revitalization, colonization." The choice of words here is important, because they are not merely actions (to regenerate, to revitalize, to colonize); they also refer to "actions of...", or even "the *activation* of...". Presenting these verbs in their noun form conjures the notion of sustained action, emphasizing the need to embed all intentions for spatial transformation in a contextual and multi-temporal foundation.

This understanding of the project as a negotiated process was gradually incorporated into the research themes proposed by Europan and is now one of their fundamental characteristics, recognized by the participating cities and accepted by the competitors.

On the one hand, as Fabien Gantois (winner of E8 and expert for Europan France[29]) explains, "when cities approach Europan, they are showing that they are looking for innovation and experimentation, that they know that the timeworn traditional methods can't adequately respond to the urban issues they face". But although this focus on the process appears to be a seductive alternative, its implementation and experimentation requires the acceptance of a certain amount of risk, which must be shared by all involved: "Europan's research themes [...] don't just require formal innovations from architect-urbanists. They require a change in the methods of urban production; a change, in turn, that calls for cities to rethink how they function, because they don't receive turnkey projects: this innovation requires them to rethink the whole 'production line'".[30] The desire for an alternative requires everyone to reexamine how they work and

29 Fabien Gantois was a project expert (project analysis committee) for sessions E10, E11, E12 and E13 and a site expert (development of site documents) for sessions E12, E13 and E14; he also carried out assessments for the post-competition follow-up for Saint-Herblain E12 and Montreuil E13.

30 Fabien Gantois, "Acclimater les projets", in *Urbanisme*, "Europan, une aventure à suivre", Special Edition n° 60, June 2017, p. 24-25.

think; all the functions and skills involved in the project must work together to invent new *ways* of transforming urban milieus. In this sense, if the project-process explored by Europan is understood by cities as the promise of multiple possible alternatives, those cities that take the plunge must also commit to undertake their own process of invention. The conditions for crafting a project-process rely on a shared effort, one that blurs the traditional lines between the client and the project manager, shifting their service-oriented relationships towards new forms of creative cooperation.

On the other hand, in view of her conviction that "an urban project is not a plan, as interesting as that may be, but rather an urban story that needs to be finished or started, and supported over the long term" Claire Schorter (expert for Europan France[31]) believes that "Europan, in focusing on approaches rather than plans, has helped to develop the methods used to create the city". She acknowledges that this notion of "process" could seem a little "banal" today, but reminds us that "this was not the case at the start of the new millennium, when urban projects were over-designed, rigid". This conception of the project as a spatio-temporal strategy has transformed contemporary urban and architectural culture, to the point that "it has been completely internalized by the younger generations, for whom it is natural to think in terms of processes, of discussion and negotiation with clients and residents"[32]. However, although these young architects

31 Claire Schorter was a project expert (project analysis committee) for sessions E8 and E9 and a site expert (development of site documents) for session E7.

32 Claire Schorter, "Des démarches plutôt que des plans", in *Urbanisme*, "Europan, une aventure à suivre", Special Edition n° 60, June 2017, p. 22-23, p. 23.

and urbanists may be steeped in this new design culture, it remains a real challenge for those who try to apply it within the context of the competition. This new *way* of project-making introduces new *types* of questions: what types of representation should we employ? What types of documents should we use to describe, define and translate these specific project dynamics into potential spaces? As explained by Carlos Arroyo (architect and member of the Europan Europe Scientific Committee), these "new types of redefinable projects" pose "a challenge in terms of their representation, and the documents that need to be produced", just as "reading the proposals [is] itself a challenge", because when these proposals arrive, "a new project language could be created"[33].

This new project language arises from the invention of new ways of narrating, new ways of story-telling, new ways of sharing our potential urban future, brought about "by a mobilization of context that is both strategic and poetic." For David Marcillon, "the practical application of 'the project-story' is the creation of a narrative thread, a matrix that has a meaning and an effectiveness, by mobilizing and linking existing situations using new innovative dynamics that operate by means of interpretation, selection, argumentation and reconfiguration".[34] As a competition of ideas, Europan was designed as a space in which to explore the poetic potential of the project output. As far

33 Carlos Arroyo, taken from his introduction to Bernd Vlay, "Quand les petits projets ont de grands effets...", in the presentation brochure for the E13 topic "The Adaptable City 2", p. 16-17, p. 16.

34 David Marcillon (architect and expert for the Europan 10 session), "Réversible/irréversible : une dualité éco-matricielle pour l'opérationnalité dc projets-récits", in *Traits urbains*, "Les fruits d'Europan 10 ", n° 39, June 2010, p. 26-30, p. 30.

back as Europan 2, the winning team[35] for the Zaanstad site in the Netherlands chose to present the central ideas of its "contextualized density" project for a strip of industrial wasteland next to the river by structuring their narrative around the metaphor of a love triangle between city, architecture and landscape. While the relationship between the first two protagonists switches between tension and complicity, the landscape insinuates itself into their story, resulting in games of seduction and other ambiguous situations; a tumultuous encounter that ends with the anticipation of a new genre of urban morphology, a "residential platform"[36]. Over the course of the recent competition sessions, while, according to Alain Maugard[37], the project-process prevailed, urban transformations were articulated via the creation of multiple scenarios (*Bondy's Count* project, Bondy, E13, see p. 155) and other micro-narratives (*Des îles partagées* project, La Corrèze, E13, see p. 56), which made the designers characters in the story they were telling (*Base Vie* project, Goussainville, E13, see p. 57). The invention of new ways of project-making can be viewed as the search for a new aesthetic, if, to use the words of Bruno Latour, "you are prepared to understand the word 'aesthetic' in terms of its ancient meaning as the ability to 'perceive' and 'relate to', in other words, the ability to be aware of oneself, an ability that precedes any distinction between the instruments of science, politics

35 Project entitled *Un rêve urbain*; winning team composed of Tania Concko and Pierre Gauthier (see p. 54-55).

36 Cf. Didier Rebois, "Questions / Hypothèses / Figures", in *Europan France 1988-2007 – Innover dialoguer réaliser*, *op. cit.*, p. 214-237, p. 226-228.

37 Extract from "Entretien avec Alain Maugard, président d'Europan France", in *Traits urbains*, supplement to n° 85, "Europan 13 : la ville adaptable, les résultats en France", p. 2.

and art".[38] The master plan has become a collective narrative. It is no longer the demiurgic visualization of a clear future; rather, it has become the foundation upon which the sharing of varied and varying interests is spatialized and recorded. For the communes of Grigny and Ris-Orangis (Europan 14), the runner-up project, *Terres Vives, les nouveaux communaux*[39] invoked the medieval history of land use (i.e., "communal" land without private ownership) to *speculate* about the forms of productive urbanity that this system could result in today; to imagine *in context* how you could organize, manage and spatialize shared land with collective rights of use, where temporary and transient activities could transform land use in response to new initiatives, whether driven by the need for new housing, arable land or facilities for education. The project did not rely on what is probable but chose to bet on what is possible. It insisted on the possibles, activated them, made them more consistent, not "just so that they could be felt, thought or imagined, but so that they would be able to make one feel, think and imagine"[40].

Imagining, representing and sharing the project process by transforming it into a narrative is another *way* of caring for the construction and transformation of urban landscapes. A new generation of architects and urbanists is maybe better suited to this attitude of sensitivity to the invariably multiple futures of places and things,

38 Bruno Latour, "L'Anthropocène et la destruction de l'image du Globe", in Émilie Hache (dir.), *De l'univers clos au monde infini*, Paris, Éditions Dehors, 2014, p. 27-54.

39 Project team composed of Maia Tüür and Yoann Dupouy (see p. 157).

40 Didier Debaise, Isabelle Stengers (dir.), *Gestes spéculatifs*, Dijon, Les Presses du Réel, coll. "Drama", 2015, p. 3-4.

to the extent that, faced with socio-economic crises and the (re)engagement with environmental challenges, it has defined its practice by its ability to deal with the uncertainty and instability of the contemporary world. However, these relation-driven and process-oriented strategies also entail a certain "risk"; the risk of forgetting "that there is a physical space to design, that that's our job, and that it's not always a matter of building, we can also work with the space in between built elements. It is our professional responsibility and we should possess the technical tools to do so." In respect of the projects for session 13, Aglaée Degros (architect-urbanist, member of the Europan Europe Scientific Committee) underscored the importance of not separating the development of strategies from space designing, of not rejecting construction in favour of processes and negotiations: on the contrary, she believes we must "find a balance between the two", in other words, we must "not forget the foundation of our discipline while we enrich it with this discussion of the right strategies to put in place."[41] We should not view these new ways of designing the urban and new ways of being designers as an alternative that has no memory or history; what we are doing is simply reworking our tools and skills, readjusting them, recalibrating them to suit each situation, intervention and implementation. Future sessions will, therefore, need to generate new reflections and experiments, and to continue to stratify them.

• • •

41 Aglaée Degros, "Concevoir l'espace et construire" (interview carried out by Antoine Loubière), in *Urbanisme*, "Europan, une aventure à suivre", Special Edition n° 60, June 2017, p. 27-28.

These four strata are formed from the synergy between saying and doing; they reflect the *in-definition at work* between words and situations. An in-definition is not the inability to define, but rather a *desire* not to finish, to fix, to settle on too quickly; that which refuses to be enclosed in certainties and prefers to open up to the plural and the possible. This first chapter aims to highlight the crucial and interactive role that this kind of in-definition has had on the themes put forward by the competition sessions, exploring the different ways in which they have reflected on the various forms of reinterpretation, in respect of each situation and in light of the possible architectural and urban responses. Next, we explore how the Europan projects work with a certain amount of in-definition too, opening themselves up to interactions between themes and cities, but also between cities and projects, and even between ideas and the collective challenge of their implementation.

CITIES THAT ARE PRODUCTIVE... INCLUSIVE AND METABOLIC?

Didier Rebois

Since its creation in 1988, Europan has always used its themes to encourage project-related innovation in response to changing of ways of life and urban practices. And while this is far from being an automatic, direct causal relationship, it opens up space for thought and action in relation to new spatial challenges – from the reinvention of mixed typologies to the creation of complex urban project-processes.

"Changing Lifestyles and Housing Architecture", the first Europan theme, asked the question: "how should we think about our home and its environment as a place for spatial innovation, in order to better adapt it to changes in the traditional family structure, the woman's role in society, new ways of working, and the focus on the individual and the body, while incorporating new technologies?". To each his own home? An ode to diversity? Or homes designed as transformable platforms? The question is still relevant today, as practices keep evolving but clients often remain extremely conventional, reproducing outdated and unsuitable forms.

"At home in the city", the third theme, encouraged strategic reflection on the porous nature of the relationship between the public space – where people are increasingly carrying out individual activities – and the private space – which posed the question of the role of homeworking. "The use of communications technology and the fragmentation of urban spaces change the notion of distance and spatial localness; social relationships are no longer confined exclusively to one's immediate surroundings or domestic life, they extend to the scale of the city." The challenge is to enable a broad range of social relationships, to encourage the use of innovative practices in the space, to manage mixed uses, and find opportunities for, and ways of, creating local jobs.

"European urbanity", the theme of Europan 8, focused on the relationship between urban practices and shared spaces at a moment of crisis for the public space. Urbanity can be defined as "a shared way of experiencing the city and its functions, but also the desire to reflect on the heterogeneity of its forms and to redesign the public space to encourage people to meet in their common spaces." However, in defining urbanity as landscape, atmosphere and place, the competition posed the question of how to interweave the (urban-architectural) scales between planning and construction; it also wanted to find the intersection of skills between architects, urbanists and landscape architects in order to produce a shared skillset that would "make urbanity more architectural".

Via the introduction, in Europan 12, of the theme of "The Adaptable City", Europan took into consideration the importance of "the environmental dimension of the urban, and the need for European cities to reduce their environmental footprint to help resolve the energy crisis, combat the greenhouse effect and preserve non-renewable resources. This transformation applies both to their morphology (form) and their metabolism (i.e. all their energy expenditure), and is highly dependent on the ways of living they propose". The theme addressed in Europan 13 introduced temporality at several levels: spaces that outlast their uses and must, therefore, be adaptable over time; urban landscapes that obey their own rules and change gradually; and project-processes that evolve, focusing less on the creation of objects and more on the metamorphosis of the existing environment, incorporating the project within multiple temporalities and rhythms, and posing the question of whether we can achieve "urbanism with less, or without, growth".

Shifting focus to productive cities, Europan sessions 14 and 15 re-interrogated the reality of diversity, a topic that had become the leitmotif of urban projects but tended to exclude the productive dimension. From artisans to forms of light industry, from research associations to manufacturers to new urban services that use cutting-edge technology – this diverse range of production methods can help to revitalize urban environments and support projects in which the concept of the urban lifestyle no longer excludes productive work. This touches on the spatial issue of compatibility between the built environment and the public space and, therefore, how to organize these spaces according to types of use. In this era, characterized by the end of functionalism and disjointed zoning, the receptiveness of cities to this theme is a sign of its relevance and perspicacity. The award-winning projects for these two sessions offered some very interesting perspectives on the potentially close relationship between making and living in the city.

However, if we are thinking about production in the city we also need to re-interrogate the relationship between nature and culture. Human activity is now a force that impacts nature and living things, to the extent that it has caused a major environmental crisis. This crisis has resulted in the need for a new and more inclusive relationship between man and nature, one that focuses more on exchange and less on domination, while also working for citizens and striving for social equality[1].

This inclusiveness refers, above all, to the inclusion of nature. We need to rethink the metabolic approach to cities, to go beyond the dualisms of nature/culture, living/non-living. Transforming our modes of urban production is one of the keys ways in which we can address the

1 See the review *Stream 04*, « Les paradoxes du vivant », 2017.

negative effects of climate change. The challenge – which is itself undergoing a revolution – is to move from a linear method that produces "lost waste" to a circular production method within a virtuous circle of environmentally-friendly production, reduced consumption and the recycling or reuse of the production "leftovers". As this circular approach develops over time, it should encourage better integration of the productive dimension into cities.

However, designing productive cities must not lead to gentrification, to mixed-use neighbourhoods reserved for the richest and those who are already well integrated. Shared productive cities are an antidote to nationalism and protectionism. Cities know that excluding living things and/or human beings is counter-productive in the long run! European cities – without being naively optimistic – were founded more on openness than on exclusion.

These themes of the inclusive and the metabolic, combined with the productive, emerged from the complex issues facing certain award-winning projects from Europan 14 and some sites from Europan 15... However, they do not need their own Europan theme, because the hybridization between them, together with the changes taking place in Europan cities, is a strategic challenge that would transform these cities into cultural AND living environments. ●

"Entre pierre et jardin"[Weak Density], E3 winning project for the Saintes site, Armand Nouvet, Thibaud Babled, Laurent Berger, Marc Reynaud and Cyrille Véran. Photos Tournebeuf.

Horizontal alleyways extend the public space into the center of the block, providing access to new private homes.

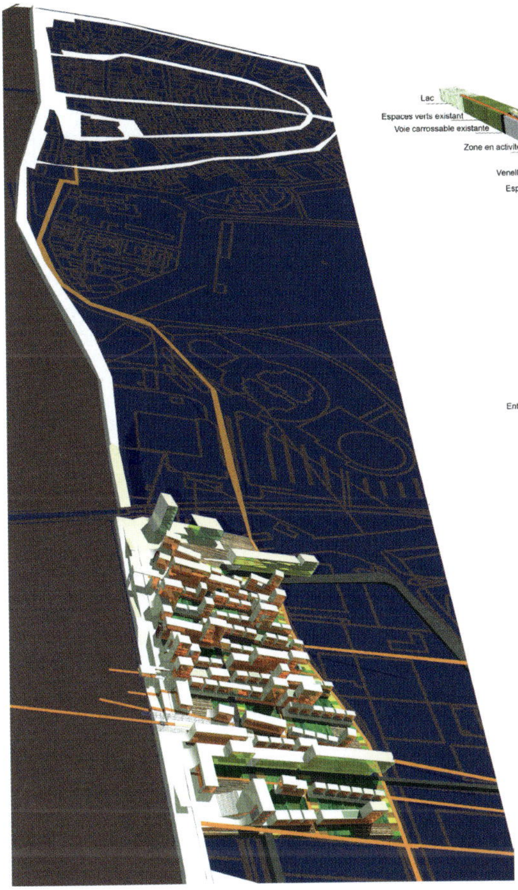

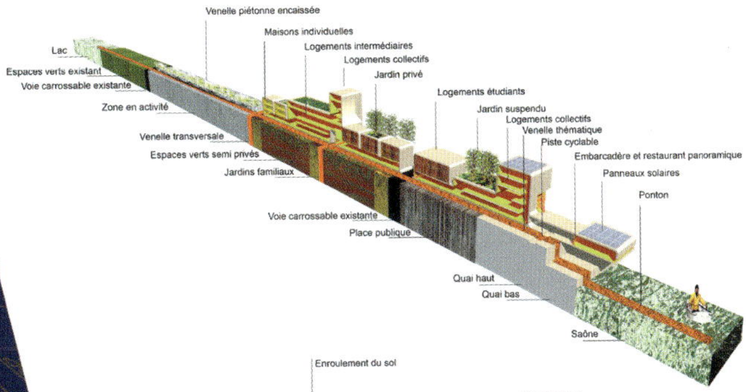

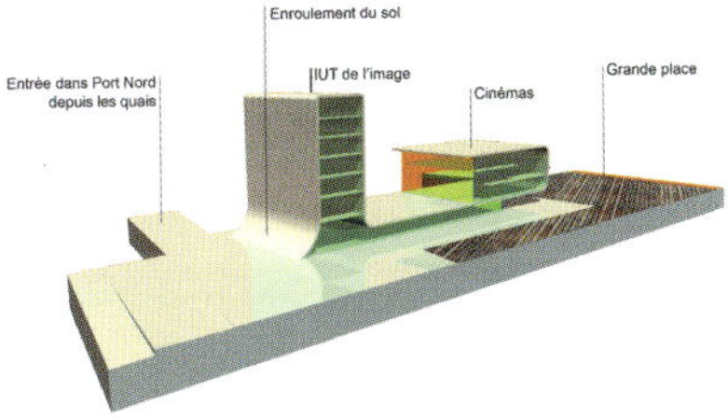

PORTE D'ENTREE

"Au fil de l'eau" [Following the stream], E8 runner-up for the Chalon-sur-Saône site, Gaëlle Cuisy, Karine Martin and Anthony Brook.

The revival of Chalon's identity as a port is achieved by means of various physical connections to the water: large ramps leading down to the river's edge punctuate low quays and provide access to boats used for tourist activities.

"La nature au quotidien" [Nature for everyday life], E8 runner-up for the Hénin Carvin site, Fabien Gantois.

The team proposed a range of interventions to bring nature into the heart of the geographical, urban and architectural devices as a programmatic, aesthetic, sensual, social and ecological element.

"Boulevard de la Mer" [Boulevard of the Sea], E13 winning project for the Saint-Brieuc site, Cyril Breton, Pierre-Olivier Carpentier and Maxime Gennevrier.

This design extended a former freight railway in order to link a range of shared projects.

"An urban dream", E2 winning project for the Zaanstad site, Tania Concko and Pierre Gautier.
Photos Rob 't Hart.

The urban plan is situated at the center of a varied and complex context; the design makes it a focal point using the surrounding canals and architectural reference points.

DOK

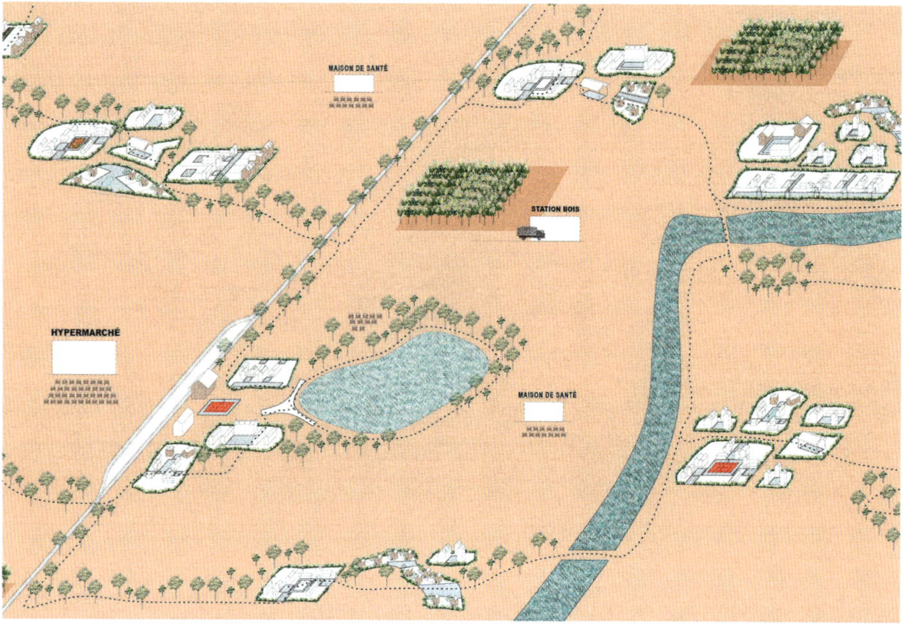

"Des îles partagées"[Sharing islands], E13 winning project for the La Corrèze site, Jeanne Moullet, Yasmine Gaizi and Victor Miot.

The flexible structure of the archipelago promotes diversity.

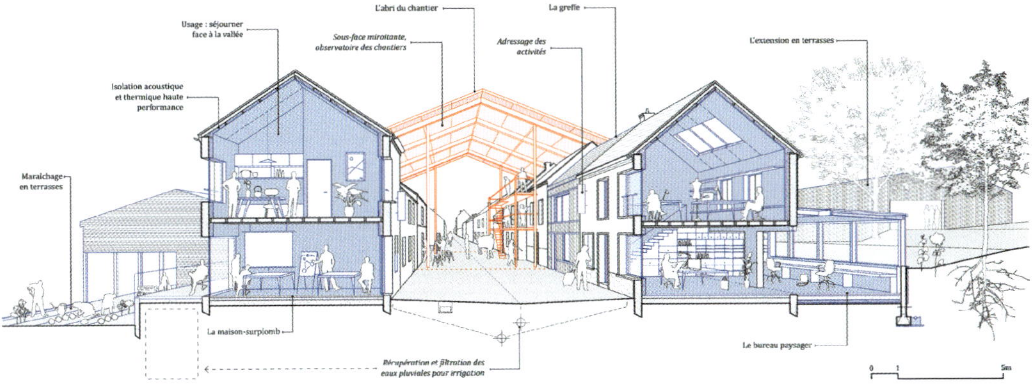

"Base Vie " [Life Base], E13 winning project for the Goussainville site, Florent Vidaling and Camille Le Bivic.

The project needed to consider temporality at several levels: it needed to start immediately but also to take a long-term view.

PART TWO

MAKING A METHOD

Chris Younès and Céline Bodart

Europan is a source goldmine of experiments and explorations that demonstrate the richness of collaborative, pragmatic and democratic approaches to the implementation of situated urban practices. These approaches recognize and mobilize specific milieus through individuals, collectives, institutions, politicians and citizens, using interventions that invent new ways of making together. A culture of open processes, sharing what their implementation puts at stake, is vital when there is the need to manage uncertainty and to offer an alternative to rigid plans by establishing rules and mediation between partners. This is how we can encourage experimentation, taking advantage of the opportunities derived from the intersection between the "top down" and "bottom up" approaches, and from horizontality or capillarity, in order to catalyse and stimulate fertile innovations.

Now, the construction of the Europan method should be viewed as the construction of collaborative, contextualized and creative projects; a method that was founded and formed collectively, between the local and the translocal, in the European melting pot. Over the course of the sessions and projects, tools and approaches were transformed and possibilities were formulated – but space was always left for new unforeseen scenarios to arise. In this sense, it is import to state from the outset that when we talk about making the Europan method, that is not to say that *all* the submitted and/or award-winning projects followed the same model. Rather, we will examine how, on the basis of their constraints, and by means of their collaborative approaches, they all participated in the construction of a new way of perceiving the project, in all its scales.

ANOTHER ECOLOGY OF ACTION

Even in tough and turbulent times, the city has never stopped recomposing and reinventing itself, in spite of more or less homogenizing urbanist intentionalities and the brutalization associated with the drive for profit and a technocratic arrogance which, instead of viewing the city as a complex and fragile living milieu requiring the utmost care, has too often objectified and abused it. In this sense, the 20th century represented the battleground between two fiercely antagonistic visions: that of the city managed by planning alone versus the spontaneous city with no design and no control. These two simplifying visions resulted in a kind of anti-city characterized by the absence of quality and by potentially explosive situations; the degradation of areas associated with an extreme inequality of living conditions and more people living in uninhabitable milieus.

However, the city's endurance across space and time certainly speaks to its ability to transform and to handle social interactions and existential stimulation. In order to "make a city", we need to become one with it, to get caught up in its rhythms, both individually and with others, but also to define the conditions for acting and living together. It's not enough to denounce root causes such as the harm caused by inequality, exclusion and pollution. In order to create the territory, we need to develop new *ways*, not styles, of seeing, thinking and building in terms of the relationship between nature and culture, between local and global, and between professional skills, but also between individual desires.

Therein lies the challenge of our present world and its possible becomings; it is this challenge that requires urban designers to expand their efforts and their imaginations in order to invent and construct new project attitudes and methods, rejecting the temptation to engage in causal finger-pointing in favour of re-learning how to think about and work with the consequences, and, above all, to bring to these consequences a certain power of assemblage. However, rethinking how we are used to making the city and changing our attitudes in order to inherit a past vision requires a certain amount of *daring*: daring to go beyond and bypass approaches that have become "the standard"; daring to transgress the limits imposed by methods that are no longer questioned; daring to confront all the complexity of inhabited milieus – which interweave the physical, cultural and social information of the territory and its technocratic rules – and responding with urban, territorial and architectural projects capable of engaging new potential and desirable forms of becoming. Daring to experiment, to explore new urban imaginaries driven by the idea that "there is nothing more important today than being able to collectively invent tools to protect ourselves from despair and cynicism, such as words that suspend the normal course of things and (re) create the possible."[1] It is such a design-attitude that we propose to revisit here as an underlying theme of many award-winning Europan projects.

As a competition of ideas, Europan addressed new generations of architects, urbanists and landscape architects, inviting them to dare to imagine how their practices might take part in the invention of an alternative making of the inhabited milieus. In addition, as a competition of ideas with *implementation*, Europan pushed designers to test their idea-driven projects against the real, to build and support the necessary situated *and* shared conditions of the work of the imagination. Put in these terms, it is clear that what is needed is a new understanding of the idea of "imagination" itself. According to the anthropologist Arjun Appadurai, it's no longer enough to consider imagination as simply the product of private or individual fantasy, expressed as an escape from reality. Rather, it is the ability of imagination to become collective that expresses its agency ("puissance d'agir"): "The imagination is today a staging ground for action, and not only for escape."[2] It is when they are shared that the ideas imagined within

1 Émilie Hache, "Where the Future is", preface to Starhawk, *Rêver l'obscur : femmes, magies et politique* [1982], Paris, Cambourakis, 2015, p. 7-25, p. 20.

2 Arjun Appadurai, *Modernity at large : Cultural dimensions of globalization*, Minneapolis, University of Minnesota Press, 1996, p. 7. This redefinition of the work of the imagination had already been used and reinterpreted in terms of architecture projects by Cynthia Davidson and Monica Ponce de Leon, in their exhibition entitled "The Architectural Imagination" for the United States of America pavilion at the Venice Biennale in 2016. Cf. Cynthia Davidson, "The Architectural Imagination", in *LOG 37 : The Architectural Imagination Catalog*, June 2016.

a project become common ground, a "prelude" to the staging and implementation of the various architectural and urban scenarios that will be selected and negotiated in context.

In this sense, we need to examine how the Europan competition has enabled the redesign of, and encouraged experimentation with, the urban transformation project as a *collective* work of imagination, capable of engaging, in each location, an alternative ecology of action. This collective dimension must be tracked here in several stages: first, when the project site representatives address Europan and its ability to translate an urban problem situation into a competitive and themed call for ideas; then, when the award-winning projects encounter the site and the agents of their potential implementation; and, finally, when each experimentation location is invited to take part in a Europe-wide debate, generating a genuine network for urban reflection and experimentation, a creative and trans-cultural ecosystem. It is via the interweaving of these three specific moments of sharing that Europan and its projects aim to "make a method", and that is what we aim to observe and summarize here. In the context of a constant back and forth between the different skills involved in the project, between the initial ideas and their potential futures, how do we elaborate a different method of thinking, designing and acting *collectively* with the ongoing transformations of our inhabited milieus?

A SITUATED AND CO-CONSTRUCTED PROBLEM

In the beginning is the milieu – a French word meaning both "the middle" and "the environment" (Umwelt). The acknowledgment of a milieu inhabited by multiple human and non-human entities; a milieu enlivened by various interests and non-interests, but above all a milieu that is troubled, affected, concerned with what has happened to it, what is happening to it and what could happen to it. In other words, everything begins when a local authority approaches Europan with an urban site that is causing it problems; spatial configurations that are disturbing and troubling the vision of its potential future. As Didier Rebois explains, "a site put forward by Europan is not a program or a space that has been planned and needs to be formalized. It is an *urban problem*, a series of often paradoxical elements inherited from past waves of urbanization. The challenge of this competition is, above all, these spatial contradictions, this discordance between the territorial, the urban and the architectural scales."[3] Although this initial problem is translated in spatial terms, it's important to remember that it is, above all, individuals who express its importance; regional planning officers or managers lending their voice to what is posing a problem for the city, choosing to make it count, to give importance to the exploration of its possible futures. What is needed is a milieu capable of being *concerned* with the social and political conditions of its spatial and physical transformation.

3 Didier Rebois, "Questions, Hypothèses, Figures", *op. cit.*, p. 219. Emphasis ours.

In describing the initial process of the Europan competition in this way, we are explicitly reinterpreting its key features through the political theories of the American philosopher John Dewey. In *The Public and its Problems*[4], Dewey proposes that we reconsider the notion of experience as a *continuum*, an "interaction" between feeling and acting; between the individuals who experience an issue and their ability to call a public into existence in order to respond to what affects them and, therefore, also initiate and take part in possible actions. "The public" is not a formless authority, not a pre-existent or pre-determined structure. As Dewey sees it, we need a specific experimental framework, a *tool* for experimentation (which he calls "inquiry") so that a public comes into existence, forming itself through the acknowledgement of a problem situation, and of the needs and interests involved. Our hypothesis is that Europan could be considered such a tool. A milieu that is facing an issue will find in Europan the means to formulate, express and translate this into a territorial and urban question[5]. Supported by their national secretariats and via a process of back and forth with the themes for each competition session, local authorities can

4 John Dewey, *Le Public et ses problèmes* [1927], Paris, Gallimard, coll. "Folio/Essais", 2010.

5 Although it is often local authorities that contact Europan France, other "go-betweens" may also be involved to facilitate the interaction (planning agencies, councils for architecture, town planning and the environment (Conseils d'architecture, d'urbanisme et d'environnement or caue), government architects or urban planners, public land agencies (Établissements publics fonciers or epf), public planning bodies (Établissements publics d'Aménagement, or epa), or even French national schools of architecture). Europan France also approaches territories prospectively, based on thc themes of the competitions announced.

transform a problematic situation into a project site. They tackle the issue of the site's future by sharing it, allowing others to share in the task of imagining and making new stories for its future.

The decision by a local authority to call on Europan marks the first step in a collaborative approach. It is the start of a transformative process that begins with an experienced trouble, a dispersed problem which is felt and named, and then engages in co-construction of a situated and shared problematic. To insist on the "dispersed" nature of the problem initially encountered, is to feature it as always-already there, present in a form which is scattered and dispersed, and thus precisely calling for the invention of new assemblages. Europan, with its highly specific competition methods and processes, is one way to meet this call.

To understand how Europan works as this kind of tool for assemblage, it needs to look at how each national structure supports local authorities to define their project site but also to outline the challenges and expectations for the client, and always in resonance with the theme of the particular session. For example, Europan France organizes a "Site orientation seminar", bringing together the representatives from all the national sites for a given session, in addition to members of the jury and a range of guests invited to enrich and energize the dialogue with local authorities, in order to refine and re-define the challenges of the project sites. In addition, an expert is assigned to each project site by Europan France to foster and maintain links with the local authority on the ground, from drafting the site files to address to the competitors, to providing support to the selected team involved in an implementation process. It is important to mobilize experts but also to choose local public and/or private partners, as their presence and advice from the very beginning of the competition session strengthen the process as a whole, lending it stability in the face of the multiple and inevitable turbulences of any architectural and urban project. Thus, a community is formed around a transformation project; a *public* that co-constructs the conditions to enable it to act on, or more accurately, *interact with*, an urban situation that has been experienced and acknowledged as problematic.

Presenting this time of work as a collective and intertwining moment between Europan and the local authorities, aims at comprehending what holds them together beyond the sole funded order. It underscores how, with the recognition of their respective and reciprocal bonds, both undergo a necessary and mutual transformation. Although site representatives often speak about[6] what they gained from competition in terms of project possibilities – whether in the form of big or small surprises and/or disappointments –, they just as often mention the shared experience. The collective aspect of the whole process is expressed through its ability to question and put into perspective the usual ways we think about and carry out the work of urban transformations. As Pierre Barros, Mayor of the city of Fosses (Europan 12) put it: "Europan is a meeting ground". A meeting ground that forms and transforms each one, but, above all, a meeting ground that requires everyone to "devote time", to re-engage with the importance of "taking time"; taking time to work with inhabitants, time to raise awareness and mobilize interests of local partners, time to discuss and debate with other cities and other local and trans-local stakeholders:

6 We also reference the "Regards croisés" sections in the Special Editions of the journal *Traits urbains*, published following the announcement of the results for each competition session.

"we dedicated time to it. But this work is fundamental to understanding how you develop the city. It's a great opportunity to work and talk with city specialists but also to meet other cities that, like us, have a territory they want to preserve by opening it up as much as possible to alternative solutions and by getting off from the beaten track of urban renewal. It allows us to sidestep the more traditional methods."[7] What the competition *makes* goes far beyond just its results. That which it transforms cannot be reduced simply to its potential physical interventions in a given territory, because, through the work it does, and for those it engages and binds along the way, it can have multiple effects that are just as important and long-lasting. Between the project site, its design-thinking framework and its implementation conditions; between representatives of commission, its initiators[8], actors and mediators: all are co-agents of their own transformations.

7 Pierre Barros, taken from "Regards croisés des représentants de site", in *Traits urbains*, n° 72, "Europan 12 France : la ville adaptable", 2014, p. 9-16, p. 11.

8 In another field of experimentation, this notion of reshuffling the deck in terms of the commission draws on the Protocol of the New Patrons, established by the artist François Hers. Cf. *Faire art comme on fait société – Les Nouveaux commanditaires* (edited by Didier Debaise, Xavier Douroux, Christian Joschke, Anne Pontégnie and Katrin Solhdju), Dijon, Les presses du réel, 2013.

AN ENGAGEMENT WITH IDEAS, CONTEXT AND PROCESS

So the definition of an Europan project site depends on the construction of a collective concerned with transformation conditions of a specific situation. But such a collective dimension, created *to* host project ideas, must also allow itself to be transformed by that which it hosts. This is the second stage during which the becoming of a site is defined by its capacity to be shared, this time between its representatives and the transformation scenarios projected onto it by the competitors.

The moment when the award-winning projects encounter the site and the agents of their possible implementation is a decisive moment of "transition" for Europan: "The implementation phase has a transitory quality, it is a rite of passage from one state to another. The initial state, the competition phase, is the construction of "images of the possible", the enrichment of an imagined future of urban life by the people involved in the project. The final state, the state to which this transition leads, is the transformation of the built environment."[9] In a certain sense, placing all the importance of the competition on the stage that comes *after* the competition seems paradoxical. However, we need to remember that the Europan competition is not just some kind of ideas-machine for urban design at a European level, intended to be some kind of generator of urban ideas at

9 Socrates Stratis, "Bienvenue encore une fois dans mon jardin" : une interrogation sur la porosité urbaine", in Didier Rebois *et al.*, *Idées en mutation : réalisations Europan 2008-2012*, Paris, Europan Europe, 2012, p. 46-50, p. 47.

a European level, satisfied with the "paper project" as the only purpose of its mission statement: the curtains closing once the winning projects have been announced. On the contrary, Europan's engagement goes far beyond the competition period. Its various bodies take on an important *mediation* role, in order to provide concrete follow-ups to what has been produced during the competition process: that means to organize public meetings and exhibitions, but also workshops and other kinds of collective working events, once again based upon the will to ensure a better support through this delicate transition time in which projects and sites first encounter one another. After the competition, Europan helps to compose a milieu for the sharing of ideas: a *prelude* to their implementation.

This commitment was not, however, always one of the competition's goals. Although today it is one of its key features, it actually evolved and was formalized gradually, as the sessions progressed. The competition learns from and co-evolves with what it produces, to the extent that this mediation role has become one of the central missions of the current Europan structure.

The transition from award-winning idea to implementation is as important as it is difficult and demanding. It is at this moment that the project's good intentions come up against the constraints of reality, the site's local dynamics and sensitivities, its network of interests, and other limitations introduced by a myriad of inevitable habits of thinking about and of acting *on* (rather than *with*) the built environment. It is the moment where the words and images of the competition project are put to the test of a situated concreteness. So, how can ideas intended to ensure the "fluidity" and "porosity" of the regeneration of inhabited milieus withstand the implacable logic of the privacy/safety dynamic? What operational frameworks need to be invented to achieve "new mixed functions and uses", and what technical and political support systems are needed to implement "hybrid" urban programs and other "typological experiments"? The development of an "alternative urbanity" that engages "new forms of interaction with the living milieu" and introduces the possibility of "new kinds of shared space" is directly conditional upon a certain change in behaviour; but how can everyone be encouraged to take part in such a change, those who side with decision-makers and developers, as well as those who side with managers and users?

Working together on these many questions requires the initiation of a *negotiation* period, that is the implementation of "a negotiated project process". In their introductory article to one of the first books dedicated to implementation processes of Europan projects (2008-2012), Didier Rebois and Stéphane Bonzani sketch "an "ideal scenario" – or rather "an idea-driven scenario" of how a negotiation could unfold. This generic scenario defines a negotiation timescale, over which the various phases will take place. The intention is not to prescribe a "template", but rather to provide an opportunity to identify the critical points in the negotiation; both sticking and tipping points, that refine what is at stake and define how is distributed the roles and responsibilities for each of the partners involved."[10] This scenario, addressed as both "a tool for thinking on Europan projects achievements" and as "an instrument for working on the future of the competition itself", is made up of three sequences of negotiations. The first refers to the "preparation of the framework of the commission associated with the site"; a sequence via which the national Europan structures and

10 Didier Rebois, Stéphane Bonzani, "Entre processus idéal et réalités" (introduction), in *Idées en mutation : réalisations Europan 2008-2012*, *ibid.*, p. 8-13, p. 9.

site representatives negotiate the formulation of a situated urban question in a way that relates to the theme of the relevant session. This is a fundamental preliminary stage, inasmuch as its degree of definition has a direct impact on subsequent negotiations, as well as the possible future achievements. The second sequence corresponds to "a communication phase"; a period in which the award-winning ideas are shared and (re-)appropriated, using various forms of mediation and discussion to enable the different parties involved in the project to become acquainted with them. It is a period of accul turation during which those involved negotiate the terms of a potential collaboration to develop and rework the ideas submitted to the competition. According to this "idea-driven scenario", the negotiations carried out in the mediation phase should also act as the impetus for the third and final "realisation" sequence. The aim of this phase is to lead to an implementation contract, committing the award-winning teams to deliver one or more parts of their projects as renegotiated with the local partners. However, although implementation is presented as both a prolongation and completion the ideas elaborated in the context of the competition, it is also in the transition towards this last step that all the work undertaken in the previous phases is exposed to the greatest risk, as the project realisation mainly relies on the (increasingly) private and financial interests of developers that are (too often) kept outside the negotiated project process. This critical articulation calls for the attention and efforts of each of the parties: "It's during this phase that the baton is passed from local strategic urban governance to the public or private developer. The city should pay particular attention to this transition as, in order to ensure the project's success, it must, as far as possible, remain a partner until the end, safeguarding the intention of the original urban concepts and continuing to monitor their operationality. Europan's national structures are also responsible for this task, which begins in the previous phases, and which ensures the quality of the realisation phase."[11] In other words, all the negotiation sequences, from the definition of the project site to the choice of building contractor, and all the partners (a broad spectrum that includes policy-makers, local technical services, planners and developers, Europan's national and trans-national structures, the winning teams and their potential associated organizations, inhabitants and other users) condition the potential physical transformations of the site.

But how can the ideas arising from the competition project survive these multiple rounds of negotiation? How can the originality and daring of an award-winning project resist, persist, potentially densify and intensify, in the face of these various meetings and interests (even when these are driven by the same desire to translate a transformation scenario into a built space)? What transforms and/or remains the same? What is won, lost and/or conceded as the theoretical project is shared and becomes collective?

The opportunity to put ideas to the test is what makes Europan "fundamentally didactic": "In the Europan school, architects learn about the complex network of political stakeholders involved in urban decisions, about the art of arguing for their ideas, but also about the need to negotiate and make concessions."[12] Moreover, in taking a cross-cutting view of these various "test" situations made possible by Europan, it appears that a unique

11 *Ibid.*, p. 10

12 Rafaël Magrou, "Jeux d'acteurs et négociations", in *Europan France 1988-2007 – Innover dialoguer réaliser*, Paris, Jean-Michel Place, 2007, p. 185-200, p. 198.

genre of project tends to be formed. This particular genre of architectural and urban project begins with a bold idea, in other words, a hypothesis of transformation that is capable of reinventing how we see and appreciate a given context, but also produces a flexible implementation strategy that keeps its potential physical representation open and partially undefined. It is at the intersection of these three elements (the bold idea, the attention to context, and the open process) and through their synergies that is crafted a project capable of holding negotiations between the imagination and the real.

THE IDEA – THAT WHICH GIVES THE NEW IMPULSE

Having an "original" and "innovative" idea is one thing, but being able to *convey* it is another. The competition format, with its presentation boards and statements of intent, forces teams to invent new modes of representation to match the originality of their ideas. The ability to translate the uniqueness of an idea into drawings and models has a significant influence over the whole project process, from the moment it is received and evaluated by the jury, to the open negotiations regarding its implementation.

During the jury deliberations[13], it is well and truly the ideas that are rewarded. Carried out in two rounds, the jury takes place in the presence of the site representatives, but the jury takes its decisions in complete independence. This seems a small and subtle detail in the

13 Submissions are anonymous, and the number may vary from fifteen to fifty for each site. For each site, the jury selects three teams (there is a winning team and a runner-up, and an additional team may receive a special mention) which will then be invited to meet the various representatives (local authorities and partners).

whole competition procedure, and yet it is what makes its very originality and creative strength. Europan does not reward projects because they provide an efficient response to a predetermined commission; rather, the competition results tend to put forward projects that dare to restate issues addressed by the initial competition brief, that push the boundaries, that reinterpret constraints and opportunities for a local context, but also in view of a broader urban and architectural culture that extends to trans-local stakes. In the competition process, the jury deliberations are, above all, a time for lively debate, and the project images become the focal point for the many and varied interests involved. Herein lies the strategic value of a project's chosen modes of representation: what form or forms should be used to represent an idea so that it speaks to *multiple* skill sets? So that it can relate to and stimulate debate between more than one understanding of the urban-architectural project? These questions are and remain necessarily open; constantly revisited in light of new project situations and new ideas contributed by each designer. For each idea, the question of modes of representation is a matter of plurality. However, this plurality should not be misunderstood as an "everything can be done", wrongly associated with a "whatever" or "however" approach. Exploring and choosing to experiment with the plurality of modes of representation is a task that should start with what the philosopher Barbara Cassin calls a "dedicated comparative in a specific situation"[14]: if everything is not equal, it is not because one choice is *better than* the other, but rather that it is *better for* a given context. This concept of "better for", places the emphasis on the performative aspect of the modes of representation: choosing the best way of wording and

14 Barbara Cassin, *Éloge de la traduction : compliquer l'universel*, Paris, Fayard, 2016, p. 157

drawing an idea's potential; experimenting new ways of telling and seeing it, and, above all, of making tell, making feel, making see, even making make.

In this sense, the aim of the competition should not be understood as a search for "the" single spectacular, extraordinary idea expressed in a flawless representation. What drives it, rather, is the construction of a framework for urban and architectural reflection, through which a proposed idea can be received as *disruptive*; an idea that can disrupt the *status quo* of a given situation, that can renew the imagination around possible futures of a site by bringing back into play what makes it a specific spatial and social configuration. A disruptive idea has less to do with what it (re)presents on paper than the effects it could have for those to whom it is addressed. The award-winning ideas, because they blur the dividing line between the possible and the impossible, while trying to avoid the pitfalls of the unrealisable, must be regarded as *hypotheses*. These hypotheses problematize – in a new, unexpected, alternative register – the conditions and forms of the transformation of a given territory, on the basis that their validity has less to do with what they are than what they enable to make happen. During the competition, *what they are* is presented in forms of varying levels of completion. But these project-hypotheses should, above all, "be accepted as the bases for actions which test them, not as finalities"; they demand to be understood as "always open to development through *use*. [...] They are *tools*. As in the case of all tools, their value resides not in themselves but in their capacity to work shown in the consequences of their use."[15] The importance of the project-hypothesis primarily resides in the effects it is able to generate, which develop around an idea that is made shareable and, therefore, capable of *impulsing* a collective and collaborative project process.

THE CONTEXT – THAT WHICH ANCHORS THE IDEAS PROJECT

An idea is also conveyed by means of a project *attitude*; a way of expressing and transmitting, through spatial and physical representations of the possible becomings of a territory, an engagement with and a close attention to the multiple dimensions of its context, which are cultural, social, environmental, economic, political, et cetera, et cetera ... The list of all the dimensions that make up the richness of the so-called "context" is never-ending, all the more because its inexticable density and instability must be taken into account. However, to anchor the project in its context is to *apprehend it with indirectness.* It is to learn to see it from a survey paying an obsessive attention to details, and not as a slippery and elusive "whole". Resolutely tactical, this project approach requires the ability to carefully track - and not without some tricks - the singularities of a milieu, its flaws and irregularities, but also its resources, more or less latent[16], in order

15 John Dewey, *Reconstruction en philosophie* [1920], trad. P. Di Mascio, Paris, Gallimard, 2013, p. 163. We have borrowed the words of the American philosopher, and in the process have radically misappropriated his argument, as it did not pertain to the methods of the urban-architectural project, but rather to the effects of ideas, theories and systems that form the basis of philosophical discourse and practice.

16 For a more in-depth look at this new project approach that considers the specific characteristics of a place, we recommend the trilogy co-edited by Roberto D'Arienzo and Chris Younès: *Recycler l'urbain : pour une écologie des milieux habités* (Genève : MétisPresses, 2014) ; *Ressources urbaines latentes : pour un renouveau écologique des territoires* (Genève : MétisPresses, 2016) ; *Synergies urbaines : pour un métabolisme collectif des villes* (Genève : MétisPresses, 2018).

to make them source material for the project. Over the course of the competition sessions, a renewed focus on the existing was firmly affirmed: "new approaches aim to get close to things, in closer contact with places, customs and materials, with elemental, vegetal and animal worlds, with all that is already existing. This points to a desire to unobtrusively inhabit the tangible reality of situations, to overcome the excessive self-confidence of modern rationalism that claims to be sure of itself and separate from nature."[17] Usual reference points are scrambling; all models are fading. Faced with a modern conception of the world that values division and exploitation, and that causes the exhaustion of natural and human milieus, an alternative social and political paradigm is gradually taking hold; an alternative ecology of restrained development is taking shape, one that aims to rethink the conditions for the anthropisation of territories, to imagine them as new forms of multi-species cohabitation.

Today, there is an urgent need to "fight maladaptation", to use the expression of the geographer Olivier Soubeyran, specialist in new development strategies [18]. It is not a question of re-enacting the tabula rasa on the failures of a dated urbanism, but rather of being able to learn from it, to re-interrogate it and to transform it from within, by initiating situated and combined actions. How can we build knowledge, methods and tools between disciplines and interests involved, in order to leverage both our extensive experience in changing inhabited milieus and the proliferation of new kinds of practices? Architectural

17 Xavier Bonnaud (architect, coordinator of expert assessments for Europan 9 and 10), "Les contours d'une esthétique écologique", in *Traits urbains*, "Les fruits d'Europan 10 ", n° 39, June 2010, p. 31-35, p. 32.

18 Olivier Soubeyran, *Pensée aménagiste et improvisation*, Paris, Éditions des archives contemporaines, 2015.

and urban practices are being reinvented in a myriad of new forms of synergy between the existing and its futures. In addition, the ways of "making-project" are multiplying trough the many possible articulations of a new conceptual-but-grounded set: the "making-with" and the "making-together". In his book *Together: The Rituals, Pleasures and Politics of Cooperation*[19], Richard Sennett recalls that humans have an innate need to cooperate. He highlights this attribute as "the foundation of human development" but also underscores the importance of knowing how to "make with" what is preexistent through a real willingness to be situated into the ordinary daily experience. In that sense, some new project strategies stress the need to relearn how we observe and describe, the need to revive the power of the dialogic conversation: its empathy, its capacity for engagement, and the rituals that create environments of listening, innovation and negotiation.

Using the "making-with" and "making-together" approaches, the urban-architecture project anchors itself in a territory, deploying a wealth of plural and situated imagination. In this way, the project works to activate a transcultural ecosystem, demonstrating the importance of supporting collective imagination and intelligence when interweaving spatio-temporal scales, while tackling the challenge that this represents. Because, although the cooperative and collaborative effects of a project constitute valuable ways of creating new narratives, new shared experiences, and new projects, cooperation is also a battle. It requires long-term commitment, first when readjusting our ways of paying attention to disrupt our habits

19 Richard Sennett, *Ensemble : pour une éthique de la coopération* [*Together, the rituals, pleasures and politics of cooperation*, 2012], Paris, Albin Michel, 2013.

of looking at project situations, and later during its whole co-design process. This battle requires us to take seriously the differences, conflicts, disagreements, discordances, boundaries and resistances (which are exacerbated in societies dominated by neo-liberal funding) in relation to heritage, reuse and reinvention, because consensus and metamorphosis cannot be taken for granted and neither can the balance between cooperation and competition. In this way, the concern for the "with" and the "together" generates a new understanding of the project timescales, new rhythms of transformation for inhabited territories, while simultaneously reconfiguring the roles and responsibilities of each individual based on a culture of *process* rather than procedure.

THE PROCESS - THAT WHICH ASSEMBLES, IN TIME AND SPACE

Although the project-process can be described, in strategic terms, as sensitive and adaptable to the changing context, gradually assembling its potential spatial representations, it should also be viewed as having a very specific project attitude, which could be defined, not as "pragmatic", but rather as *pragmatist*. This attitude is part of an *alternative* understanding of the world, and engages an *alternative* understanding of how to intervene and interact with the state of the world: "for the pragmatist, the world is not finished, it is still *in the making*. Rather than requiring the means to be adapted to the predetermined ends, the ends must always be revisited and tested based on the means that actually exist". In other words: "the end is not 'final', it is simply 'what is in sight' – temporary and contextual. Because 'what is in sight' depends on what is possible and desirable, and whose idea can only be given by some associative circumstances."[20] In order to bring about this kind reconfiguration of the relationship between ends and means, designers need to recalibrate and readjust their traditional design tools and instruments, as well as their modes of representation, communication and implementation.

Because the projects submitted to the competition "shake or awake the commission frame in a given context, through the reinterpretation of contexts that they produce"[21], they need to define a strategy capable of allowing everyone to participate. This experimentation with new ways of project-making must engage with design methods and tools that make the project sufficiently appropriable: "Thanks to their strong ideas, but also their still quite sketchy development, the Europan hypothesis-projects open the imagination and broaden the range of potential transformations, as long as all those concerned get involved into a negotiated project process."[22] The project-making process needs that some room for manoeuvre be intentionally left; it needs that its implementation be supported while accepting that spontaneous initiatives, or at least supposedly unexpected initiatives, may still arise. With such a negotiation process, it is the whole urban production structure that needs to reset its operational habits and procedures. Such a *need*, however, can achieve little without its counterpart: *the will.*

20 Joëlle Zask, "La politique comme expérimentation", introduction to the French edition of John Dewey, *Le Public et ses problèmes, op. cit.*, p. 25 and p. 14.

21 Didier Rebois, Stéphane Bonzani, "Entre processus idéal et réalités" (introduction), in *Idées en mutation : réalisations Europan 2008-2012, op. cit.*, p. 9.

22 *Ibid.*

The challenge of the project-process is to assemble individual wills, to bring them together over time, and then to keep them together throughout the difficulties of implementation. Of course, these new project modalities aren't without their problems: over the course of the negotiations, the good intentions of some unavoidably come up against the reticence of others. Between the ones and the others, and between their hesitations and indecisions, this may often drag on; this sometimes causes complications and clashes, and most frequently this results in conflict and chaos. However, *this* remains hard to define, for everyone, everywhere. There are as many processes as there are sites and projects.

• • •

Although Europan encourages experimentation with new genres of architectural and urban projects, shaped by multiple interactions between idea, context and process, it also creates moments of sharing and exchange around these specifically situated experiences. We need to broaden our analysis to a Europan level to see how the political and pedagogical dimensions of the urban-architectural project are being reengaged, from one site to the next.

NEW WAYS OF SHARING AND CARRYING OUT THE PROJECT PROCESS

After the local authorities and Europan structures have set out a situated urban problem, and after the negotiation process has been launched with the various project partners, there is a third phase that we need to examine in order to fully appreciate the importance of the collective dimension of the competition. This is the phase in which the *multiple* project hypotheses are shared.

A competition session may produce several transformation hypotheses for each site, each with a different physical representation depending on the spatio-temporal strategies used by the teams. This openness to a broad spectrum of possibilities and surprises is fundamental, as a Europan site aims to inspire a range of ideas, rather than a narrow search for a supposed correct response or solution. Michel Fanni, representative for the Neuilly-sur-Marne site for Europan 11, said this of his experience of the competition: "We expected ideas, and crazy responses. In the end, it was the sheer number of the responses that was interesting, rather than the responses themselves."[23] If this plurality is one of the competition's assets, the challenge is to be able to retain it in what could possibily be elaborated on the site. In this regard, Europan France has

23 Remarks by Michel Fanni (Director of Research and Innovation, Epamarne-Epafrance; Neuilly-sur-Marne site representative for E11), recorded by Pierre Derrouch in "Regards croisés des représentants des villes", in *Traits urbains* supplement to n° 55, "Europan 11 France valorise les processus", 2012, p. 10-14, p. 11.

introduced an interesting procedure[24] that allows multiple hypotheses to intersect on the same site and enables all the creative and inventive daring cultivated by Europan (as a space for experimentation outside the public procurement process) to be incorporated into official public structures. In effect, following the jury's announcement of the results, the three award-winning teams are invited to meet with the local partners to redraft their project hypotheses (together and/or separately), on the ground and with the various stakeholders. For these cities, this intermediate phase gives them longer to become better acculturated to the ideas generated by the competition, thereby improving the stability of the implementation process that may be undertaken with one (or sometimes several) of the teams finally selected. For the teams, it is an opportunity to explore their hypotheses in depth by testing them out in the real world, but also by opening them up to discussion, along with other visions and ideas of how to transform the same site. However, although this period of meeting and sharing seems to have a genuine post-competition dynamic, we should not forget the importance of the resources required to implement it, or play down the many rivalries and the range of problems it can arouse.

But it is precisely because the road to collaboration is filled with obstacles that we must help one another

24 Although this procedure is now fairly common following the competition process for French sites, it was phased-in gradually starting with sessions 9 and 10. It should also be mentioned that the start of this procedure reflects the fact that most French sites are subject to public procurement and are, therefore, regulated by French public procurement law. As a result, Europan France had to "bypass" these various constraints by figuring out how to transform the results of the competition into a point of entry to an official procedure that resembles that of a competition.

to stay the course. *To help one another*: to recount and share the tumultuous trajectory of each of the explorations and experiments of the situated projects. We need to recount the stories of these projects, with their inevitable roadblocks but also their positive breakthroughs; to narrate them and to interweave them, without conflating them; to draw them closer together without actually attempting to compare them. We need to maintain their plurality, to preserve their unique features and make them count, based on the belief that it is their differences that will engage new ways of perceiving and conceiving of the future of how we build territories.

It is this very perspective that opens the European horizon of the competition. A horizon of cross-reflections, stimulated by each of its experimentation sites; a horizon that is made up of numerous points of interaction that sometimes manifest as large international fora (at which city representatives are invited, in turn, to speak with members of the jury and the winning teams), and sometimes as collective publications (in which several experts come together to analyse the results for each session, and then carry out a critical review of their implementation). All those involved agree that these interaction points produce a "wealth of mutual learning"[25] in all areas. For the representatives of Europan cities, but also for the experts called on for each session theme, the fora are "above all, necessary because they enable people to get to know one another, to absorb the urban and architectural culture of other countries, and to share experiences. Urban colonization and its impact on the natural milieu

25 Remarks by Paul Delduc in "Entretien avec Paul Delduc, directeur général de l'aménagement, du logement et de la nature (DGALN)", in *Traits urbains*, n° 85 "Europan 13 : la ville adaptable, les résultats en France", 2016, p. 6.

have totally different meanings for a Finnish site and a Barcelona suburb."[26] Although the trans-local dimension of the exchanges is welcomed because it "shows us what is happening elsewhere", this enrichment of viewpoints also requires participants to "reflect critically, as societies and their ways of life are different"[27]. Looking beyond *your own* situation, seeking to understand other places because of and in spite of their differences: this is how Europan has chosen to help build a shared culture around new ways of project-making, with regard simultaneously to its architectural, urban, territorial and landscape dimensions.

Today, Europan is an important network for reflection in terms of territorial and urban experimentation. Thanks to its many generations of winners and its history with its partner cities, thanks to its experts and their influence in the fields of research and education, thanks to its (trans)national structures and their multiple ways of sharing what each session can produce in terms of projects (fora, workshops, publications, exhibitions, etc.), the competition has established itself at a European level as "the grand laboratory of alternative projects" to use Kaye Geipel's expression.[28] It has become a method of production, dissemination, promotion and even (and not without

26 Remarks by Pascal Dayre, recorded by Élise Thinus, in "Regards croisés de membres du jury", in *Traits urbains*, n° 39, "Inventer l'urbanité", June 2010, p. 22-25, p. 25.

27 Remarks by Eric Citerne (Director of Urbanism and Reims site representative for Europan 11), recorded by Pierre Derrouch in "Regards croisés des représentants des villes", in *Traits urbains* supplement to n° 55, " Europan 11 France valorise les processus", 2012, p. 10-14, p. 11.

28 Remarks by Kaye Geipel, in her introductory speech to the round table "Essence, modernité et défis de la ville européenne", during the Forum "Europan 30 ans

risk) legitimization for the invention of (sometimes radically) different ways of making a project. As Olivier Meheux points out, "twenty-five years ago, it would have been unusual to win a competition and not go on to build anything". But times have changed. Now, Europan enables ideas to be recognized on their own merits, and to be displayed, published and widely disseminated. Somehow, it offers a unique showcase for design projects."[29]

Europan allows ideas about alternative projects to take centre stage, while simultaneously introducing new questions and challenges, issues that are still too often left unresolved and that arise when such initial and radical ideas have to be carried out. For each and all European cities, what new tools and procedures still need to be invented to overcome the legal and political, institutional and cultural hurdles that regularly block the path between an idea and its realisation? For each urban operation, how do we ensure that the technical and financial plans for the construction phase are drawn up with as much inventiveness and creativity as that used by the designers to develop their projects?[30] Drawing on their experience

d'avance", organized by Europan France, Puca (Plan Urbanisme Construction Architecture), and the Cité de l'architecture et du patrimoine museum, under the auspices of the Ministry of Culture, and the French Ministry of Territorial Cohesion and Relations with Local Authorities, Paris, Cité de l'Architecture et du Patrimoine, 13-14-15 December 2018.

29 Remarks by Olivier Meheux, recorded in a Special Edition of the journal *Architecture d'Aujourd'hui*, December 2018.

30 Question reformulated here based on the discussion led by Bernd Vlay (member of the Europan Europe Scientific Committee), as part of the "Restructuring Infrastructure" Working Group at the "E14/E15 – Productive Cities" Inter-Sessions Forum, Brussels, 12-18 November 2018.

with the competition and their respective successes and failures, European cities are asked to collectively reconsider their usual ways of carrying out a project; to re-examine their extreme heterogeneity, and even to think of ways to harmonise their rules and regulations and other techno-legal constraints. How can the ways of carrying out a project process be shared? This will be one of the key challenges facing the Europan competition over the coming years.

Through these different phases that make up Europan's collective dimension, it is important to see how this essential sharing of ideas and project hypotheses (from their elaboration to their reception and even to their realization) is always a process of construction.

Collectiveness does not come into existence by itself. *New ways of doing* are needed to enable all the various concerns and interests to come together and learn from one another. The conditions for a potential and effective exchange are constructed, with care and attention, in order to bring together disparate project cultures united by their desire to take part.

TOWARDS A CO-CREATION OF MEANING IN THE CONSTRUCTION OF TERRITORIES

In this laboratory that is Europan, constructing the collective and cultivating the collaborative are part of processes that co-create meaning. These rely on the network of subjects, as each subject generates meaning and is capable of intersubjectivity and creativity. In this sense, the competition is not a school that promotes *one single* method. Rather, re-defining itself over and over, from and with what it produces, Europan is an invitation to experiment with and share more than one way of project-making, of making-territory. *Multiple* methods that are to be constantly reinvented, from one context to another.

Despite the plurality of these methods, however, a number of priorities for shared engagement tend to emerge, such as: 1) the importance of social and physical immersion in real on-the-ground experiences rather than the *a priori* application of models; 2) the exploration of the constraints and resources unique to each particular situation; 3) the importance of interlinking that redefines the possible assemblages between micro and macro, near and far, slow and fast, long-term plans and openness to change; 4) the crafting of methods and tools that are suited to participative and collaborative processes in the context of sustainable territorial strategies. As the sessions progress and the submitted and award-winning projects continue to evolve, the common dimension of the Europan laboratory is reinforced. This engages another kind of *ethos*, in other (French) words: *une manière de se tenir au monde*, as both a way of taking hold on and caring for the world, empowering invigorating synergies to tackle the fundamental environmental, political, social, ethical and aesthetic challenges that lie at the heart of the problems we currently face in terms of existence and coexistence.

BUILDING EUROPE FROM THE GROUND UP – IT CAN BE DONE!

Xavier Bonnaud

Over the course of my involvement with Europan, through various sessions, first as a competitor and winner, then in the context of a study carried out in Belgium, and finally as a project expert for fora and juries, I have always appreciated the opportunity to participate in a process of European construction that is worthwhile and tangible and is connected to a history.

The long tradition of the European city, through its cycles of construction, expansion, destruction and reconstruction, through its morphological and patrimonial configurations and its adventures in social solidarity and democracy[1], is, in essence, the narrative thread of this continent and accounts for much of what makes it unique[2].

Given the current culture of political abjuration and isolationism, this tangible European construction, built from the ground up, has never been more important: it has strong convictions, such as the need to share and stimulate the

1 Jean-Luc Pinol (under the direction of), *Histoire de l'Europe urbaine*, Paris, Seuil, 2003.

2 Leonardo Benevolo, *La Ville dans l'histoire européenne*, Paris, Seuil, 1993. Patrick Le Galès, *Le Retour des villes européennes : Sociétés urbaines, mondialisation, gouvernement et gouvernance*, Paris, Presses de Sciences Po, 2011.

appetite for a positive European identity through projects and territories.

In fact, although Europan does not receive any European funding, it could be argued that a certain idea of Europe is being constructed, gradually, over many years, on the back of its projects (530 cities in 23 countries over 15 sessions).

Within Europan, the sites are connected to one another, beyond their particular national identities. The regular exchanges between local stakeholders, young designers, experts and juries from the four corners of the continent engage more than just urban and architectural exploration. In their ability to share and evaluate the creative predictions of each proposal (down to the detail of the built and inhabited environments), each participant, on a small scale but with a keen sense of the current reality and debate, participates in the decentralized cultural construction of Europe, in contrast to the strictly institutional, regulatory, technocratic and top-down approach to construction that is still widespread but has been overwhelmingly rejected.

This forces us to confront the legacy bequeathed to us by the 20th century: an urbanism of disconnection and a certain alienation from the world, reinforced by a cold, artificial and often imperious built environment. The architecture promoted by Europan is helping us to do this, proposing new plural, alternative visions, and opening them up to debate. It also plays an important role in terms of political construction, enabling us, hopefully, to re-engage in a better and more rewarding way with our shared European construction site. ●

EUROPAN OR ETERNAL YOUTH

Bernard Reichen

For thirty years, Europan has prepared, debated, generated and leveraged the transformation of the European city on a broad scale. Its fifteen sessions tell the unique story of an exceptional era expressed by projects that combine innovation with anticipation. Gradually, the themes and, to an even greater extent, the sites, anchored this story in the incredible diversity of European urban cultures and expressed it at all levels of thought and action, from small towns to large metropolises.

However, what is unique about Europan is that, thirty years after its creation, the winners must still be under forty years of age. It is a testament to its continued resonance with the "real city" and the great speed with which it incorporates change. There is no dogma, no template, no generational conflict. Europan is a machine built to withstand the ravages of time.

And the time that has elapsed since Europan's beginnings has been anything but linear. Today, we are at one of the great crossroads in an urban history which has, over centuries, established the values of the European city. Given the scale of the changes to the idea of "urbanity", these past thirty years were maybe nothing more than the gradual preparation for the revolution that is taking place today.

Thirty years ago, it would have been impossible to imagine the tools that the digital revolution would place at our disposal. Equally, it is difficult to predict today what effects these tools may have in thirty years' time on our thoughts, our culture and our urban practices. Today, the clash between a constrained and formatted urban world (even when sustainable development is the format) and unprecedented societal change is a major challenge.

This insecure society, focused on reformulating the customs and modes of its interpersonal relationships, is no longer interested in looking to the future. Especially not when the city presented to it gives the distinct impression that the future is being invented using the tools of the past. The first consequence of this situation is the disintegration of the notion of the urban project itself. And the first sign is the emergence of an urban-architecture labelled "reinvention". Thanks to the pendulum effect, we are also seeing the emergence of the idea that "small is beautiful" and that the "co-production" of the city (the new leitmotif of the urban project) could, by itself, create the new inhabited world we are all waiting for.

These ideas, as benevolent as they are pragmatic, are also dangerous!

This "piecemeal" reinvention, which divides the city into small or large chunks and hands most of them over to the private sector, distances it from the genius of the European city: the continuity of shared spaces. In the end, the street plans "create heritage", and a city is only appealing when its links are places in their own right. During great periods of change, the transformation of space and society has always come from the same impetus. More than ever before, finding the balance between links and places, the Grid urbanism and the urbanism of ways of life, will be our greatest challenge. What meaning will these plans have when they are located in a peri-urban space far from the traditional city layout of streets and squares? What new "geography of behaviours" needs to be invented to accompany the new urban universe that is emerging? What we are inventing today is a palimpsest of the 21th century in the form of an urbanism of literal and figurative flows.

The Europan competition has, for many years, anticipated the end of zone urbanism. It is now perfectly positioned to tackle this new era of the European city. ●

COMPETITION LAUNCH

Launch of the Europan 11 competition, February 2011.

Launch of the Europan 14 competition, February 2017.

SITES VISITS

E14 site visit to Évreux, March 2017.
Photography Oriane Couturier.

SITES VISITS

E14 site visit to Angers, March 2017.
Photography Oriane Couturier.

E14 site visit to Toulouse, April 2017.
Photography Oriane Couturier.

PROJECTS ANALYSIS AND JURY

Europan 13 jury session, October 2015,
Palais de Tokyo.

FORUM OF CITIES AND JURIES

E14 European Cities and Juries Forum, Helsinki, Finland, October 2017.

RESULTS ANNOUNCEMENT

Teams selected at Europan 3, December 1994.

Announcement of the winners of Europan 12, December 2013.

Teams selected at Europan 13,
December 2015.

INTERSESSION FORUM

E13/E14 European Intersession Forum in Badajoz, Spain, October 2016.

EXHIBITION AND DEBATES

Exhibition of the french results of Europan 13 in Bordeaux.

Photography Arthur Péquin.

IN SITU WORKSHOPS

E12 Competition Follow-up for the Marseille Plan d'Aou site, a consultation process implemented by the Arkilab team.

E13 Competition Follow-up for the Marne-La-Vallée site, on-site workshop.

E12 Competition Follow-up for the Fosses site.

SITES

The building site in Saintes, E3, by Babled and Nouvet Studio: the stones reuse.

The Building site in Vienna, E9, by Uek Studio.

PUBLICATIONS

1

2

3

4

5

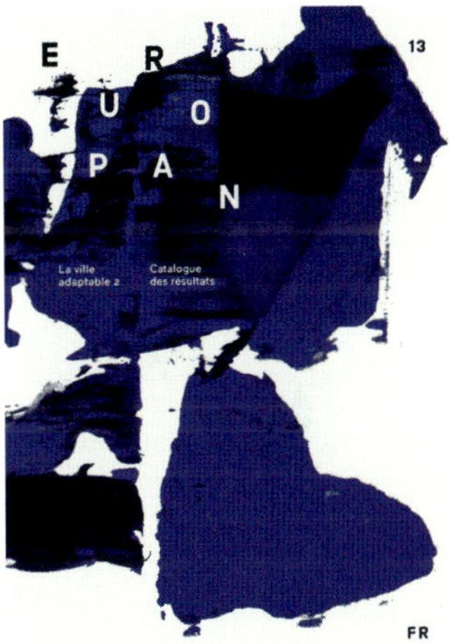

6

1 – Catalog of Europan 14 results in France.
2 – Ideas Changing, 2008-2012, implementations catalogue.
3 – Analysis of the implementations, Special Issue of the journal *Urbanisme*.
4 – Results analysis, Special Issue of *L'Architecture d'Aujourd'hui*.
5 – Monograph of the realization of Bathilde Millet in Roubaix, E6.
6 – Catalog of Europan 13 results in France.

PART THREE
TOMORROW

Conversation between Alain Maugard and Chris Younès
Interviewed by Sylvie Groueff

PREAMBLE

What challenges will architects and urbanists face in the years to come? What can they do to transform our cities? What will these professionals have in common with those working today? Change does not happen suddenly; it chips away, bit by bit, at our ways of life. Some issues finally prevail after having been ignored for a long time, such as climate change, our relationship with nature and animals, biodiversity, and living organisms in general. All these issues have already affected our ways of seeing and perceiving – such as how we think about the city and its living spaces.

The third part of this book is structured around a lively discussion between two intersecting opinions, that of Alain Maugard, an engineer and President of Europan France and Chris Younès, anthropo-philosopher and a member of the Europan Scientific Committee. Over the course of several hours, they discussed their views on the city and the urban. They touched on the opportunities created by the Europan competition and its characteristic action-project to act on the urban. Putting their observations into perspective, Chris Younès and Alain Maugard also examined how the role of architect is evolving in light of the already significant changes taking place in the profession.

TOMORROW: THE CITY

THE CITY: AS A CENTRAL PERSON

Chris Younès – The city is a matter of fact. It colonizes, transforms, undergoes metamorphosis, blurs its boundaries, sprawls. The form and content of the classical city is no longer recognizable. We are faced with a new form of urban settlement that we no longer know how to deal with. The city in its current form seems to be fading away. If everything is urban, what does it mean to be urban? What does it mean to build an urban society? Little by little the city has moved away from being enclosed, delimited, protected by its boundaries, towards becoming a city-landscape, an artefactual environment that does not follow recognizable principles. It challenges and questions.
Alain Maugard – There certainly are important questions inherent in the city itself. Is the city able to change its form, its composition? Where does the city stop? Where does it start? Why does it have no limits? Because the city essentially consists of territories that are occupied by an urban population. Urban civilization. What is a village flooded with urban immigrants? A rural village? A piece of the city? It is a territory under the influence of the urban. Michel Serres says he lives in Paris, Brussels and

in the university where he teaches in the Unites States of America. These pieces of places constitute "his" city. And there are lots of people who live in more than one city.

Isn't this object, "the city", vanishing? The more you discuss its existence, if I may say so, the more it exists as a force, a political force, an economic force. Across the world, it's cities, not States, that are planning on implementing solutions to fix our problems (global warming, COP21).

The latest Europan theme, "Productive cities", highlights the growing economic power of cities to combat globalization and its excesses. Although the city may seem to be disappearing because of its lack of limits, its existence and its force are still very tangible.

CY – The force of the city is a power in itself, a dynamic – a *dynamis* as Aristotle would call it – , which endures even if it transforms, while remaining a component part of civic, urban, human existence. In 1994, at the exhibition "La ville: Art et architecture en Europe 1870-1993 (The City - Art and Architecture in Europe, 1870-1993)", Françoise Choay, in her introduction to the catalogue, spoke about the death of the city, and pointed the finger at urbanism. She blamed this metamorphosis, which, according to her, required us to change our vocabulary to take into account an entity that no longer had the same boundaries, and certainly not the same principles of organization, but which was transforming our territories and impacting the evolution of human society.

AM – It's like saying that the city is a person and that it is changing so much that the keys to unlock and understand this person are different. And maybe this is how we can build a range of potential solutions, from the starting point of getting to know this individual, taking a kind of humanist view of this person, the city.

CY– The city is effectively a kind of person, someone who is both known and unknown, to whom we remain very close, someone who is very familiar, but not "ours". We don't go round it, we don't have any influence over it. The city doesn't adhere to some kind of dualist vision in which it either does everything by itself or, conversely, is totally controlled. The city of the 21st century forms part of the long history of humanity and the urban, like a living being that we need to understand better, to respect more. We need to view it as a complex ecosystem of which we have fragmented understanding and that science can help us to comprehend, to some extent. The city is a milieu made from interrelationships and coevolution, which revitalize the great adventures that are rural and urban human settlements. Treating the city as a living organism means looking at it in a totally different light. Indeed, Edgar Morin prefers to speak about metamorphosis rather than revolution in terms of what is happening at the moment. It certainly is a metamorphosis to consider the city a living organism, which, according to him, in the history of ideas, coincides with the ascendancy of the environmental question.

We cannot think about architecture without thinking about the city and we cannot (any longer?) think about the city without resituating it within our environmental reality. In very broad terms, we can identify three key strata in the environment-driven changes in representation brought to light by Europan. First came the stratum of the living organism: everything is connected and living things depend on one another. We need to pay closer attention to this reality of living things as a whole. Then came the stratum of sustainable development as a planetary contract, which revived the question of this living milieu, which is clearly connected to other elements. And now, the stratum of the Anthropocene reminds us that the city, as a milieu of interrelationships, is also part of the environment of the Earth. Ecology, sustainable development, and the Anthropocene are three interconnected ways of understanding this political, social and metabolic urban entity. We are going to need new approaches, new expertise and new skills if we want to be able to operate in harmony with the cycles of nature. Because when we talk about living things we must accept the cycles of life: birth, transformation and death. The exploration of the idea of the resilient city comes from this ecology of inhabited milieus.

THE CITY PARTICIPATES IN THE LIVING WORLD

AM – In the context of a more naturalist approach, I think it is helpful to think about the city using comparisons with morphology, metabolism, the brain and intelligence. Urbanism and architecture have always been very interested in the idea of morphology. There has been little exploration of and reflection on the idea of metabolism, however, and the concepts of the brain, intelligence and sensitivity have not been addressed at all. In session E13, for the site in Moulins, one team (Altitude 35, Clara Loukkal, urbanist and landscape architect, and Benoît Barnoud, architect and landscape architect) submitted a project called: "La Théorie de l'évolution" (The Theory of Evolution). The city is constantly renewing; change is permanent. Some changes, the ones adapted to inhabitants' needs, endure and are accentuated. Others are less suitable and, unless they disappear on their own, must be extinguished. This could be considered a form of intensified Darwinism, with phases of emerging change, phases of renewal, and phases of acceleration – such as that instigated by the arrival of digital technology. Europan, in the themes it proposes, has always recognized the need to account for the constantly changing nature of the city. The Darwinian thesis takes the form of local mutations. We will see which ones last and which ones don't work. Meanwhile, the new big thing in the field of living organisms is epigenetics, which is an extension of genetics: we know an organism passes on half of its genes to its offspring and Darwinian evolution acts on the combination of genes from the two different parents to create the genetic makeup of the individual. However, epigenetics has discovered that we don't use all of our genes equally. Let's think of our genes as keys on a piano. Epigenetics determines how hard or softly the keys are pressed, and whether the notes are short or long. It results from our acquired characteristics, our environment, our behaviour, our education, etc.

So, what is its relationship to the city? It has been shown that we pass on a part of this epigenetics to our children. This is essentially what Lamarck predicted. We not only pass on a selection of musical notes to our offspring, but also the way in which the keys on the piano are pressed. We are currently learning to what extent living things are Darwinian or Lamarckian. As the city is undergoing constant regeneration, there's no reason why its acquired characteristics couldn't also be transmitted: its turbulent or peaceful nature, its appearance, its "way of life". All the characteristics acquired through the social organization of the city are passed on too. Its genetics, its genes, are its streets. But the function I choose for this street, that's epigenetics, that's the way in which I press the piano keys. It could be a road for cars, a motorway, a riverside road. If I decide to turn this street into a beach, or to pedestrianize it, I am composing, I am altering its genetic morphology. It may be a little transgressive to apply the theories of Darwin and Lamarck to the evolution of the city, but the Europan candidates who have been brave enough to tackle the issues in this way have given us much food for thought.

CY– This focus on the interaction between an organism and its environment is a significant contribution from the naturalists of the 19th century. Today, architects often use the term "the DNA of an environment". What about the DNA of the environment of the city, both in terms of what is visible and invisible? This also supposes that an environment has a conscious and an unconscious dimension. It is a kind of "retroprospective". This power is returned to the place, and to the interaction of the organism with it. Relating this question to epigenetics is valuable for all those who work on the topic of human settlements, because it approaches the built object as part of a larger whole that is co-evolving. Lamarck and Darwin were the first to try to develop scientific hypotheses of coevolution. Epigenetics explores how an organism coevolves with its environment. The environmental conditions are internalized, and this internal-external relationship with an organism is particularly relevant for architecture and the city. So, how do we establish the internal-external relationship? How do human settlements put in place systems to protect that which is internal, as well as the connections to the outside? These questions represent important new areas of research.

At the dawn of the culture of this 21st century, we should reread Darwin and pay attention to what he actually wrote. Darwin didn't just examine competition between organisms; he also spoke of cooperation between species, as do many other works which investigate these forms of interdependence. The recent bestselling book from German writer and forester Peter Wohlleben *The Hidden Life of Trees: What they Feel, How they Communicate* puts forward the hypothesis that forests are cooperative living milieu. He describes the systems that trees use to communicate and cooperate with one another. We have placed too much importance on the role of competition in adaptation. Cooperation also has its place. And the city, like all human settlements, is, first and foremost, a system for cooperation.

CITIES ARE BECOMING MORE COMPLEX, LIKE LIVING ORGANISMS

AM– I actually think we need to move away from this individualism that places everyone in competition with everyone else. We need to focus on the fact that the city is a place of cooperation and collective power. The example of the forest is particularly relevant for our fields. Reading Wohlleben, I discovered that trees have a memory, that they feel pain. The roots of trees communicate with one another. The biggest tree, which has a vast network of roots, is the one that absorbs all the sunlight. It looks after the weaker trees around it. But when we talk about Darwinism in relation to the city, what levels of cooperation are there? It is the collective organization of the city that provides its reactive power, its ability to adapt to the environment. This comparison with living organisms impacts how we act on the city. We don't interact in the same way with a living body as we do with an inert one. From the moment we view the city as a living body, the internal and the external become vitally important. This relationship was already inherent in architectural thought in terms of the penetration of light from the outside to the inside. When architects started designing kinetic facades, classical architecture was transformed. The shifts in opacity and transparency lent architecture a sense of movement. Light is a vector of design and, as a result, opens up avenues for adaptation that can be extended to all levels (the building, the neighbourhood, the city). The city, a living being, has an enormous capacity for evolution and transformation. It can, in extreme situations, self-destruct or, conversely, evolve towards a better future.

The desire to be in contact with nature, which some consider naive – oh, your vegetalized facades, your fifth facades with their green roofs –, corresponds to a fundamental need for balance which the city should provide.

CY– In many Europan projects, and this is probably going to increase, the candidates internalized the coexistence between humans, plants and animals. For a long time, in Western society, humans tried to build a milieu for themselves to rule over, whereas now, living beings in all their forms are taken seriously, as is the search for symbiosis. It's a huge shift in the way we think about the urban question – and the way we approach the city as a metabolism. It's clear that a binary opposition between metabolism and politics would be nonsense. The city of tomorrow, the city of today, and the city of the past are all profoundly political. The city has the means to act collectively, and to exist as a politically driven collective milieu. Hannah Arendt demonstrated that the city is the political space par excellence; in terms of its capacity for coexistence and its ability to govern groups of people. In order to create synergies from the cycles and metabolisms of living beings we need to change our behaviours and expand our knowledge and techniques. We need to rediscover how to re-work, to repair, to recycle. It's political.

AM– What we mean by the living reality of the city is that a politics, a political governance, should be built upon this organic complexity. For some, building a city just means constructing bridges, roads, infrastructure, as if it were relevant to focus on the skeleton and morphology alone. These days, it is equally important to act on the city's metabolism, with an understanding of how the living organism functions. And why not act on its moods, emotions, and feelings too? If we push the comparison between cities and living things a step further, we need to address their illnesses, viruses. Indeed, some speak of an urban virus. In other words, we will never be able to stop working to keep the city alive. It could be subject to attack, there may be battles. This living organism may be forced to create antibodies. A resilient city is one that reacts to protect its health. Therefore, architects and urbanists must start to think like doctors and surgeons, in order to work on the morphology, the metabolism and the brain of the city, and even (why not?) its psychology.

CY– It's surprising to rediscover this medical-sounding vocabulary, which has been rejected for seeming overly organic and therapeutic. When it comes to the living conditions of humans, plants and animals, we must start from the position that, even if we aren't experts in the "whole", everything is connected and should be maintained; taken care of. In terms of the Europan entries, some teams, inspired by other well-known architects, explained their approaches by describing the architect as an acupuncturist. Performing urban acupuncture is, in a way, about working with energies, working on sensitive points that have an effect on the whole. Even if you can't treat the "whole". For example, there is a lot of work being done at the moment on new forms of energy. A living being cannot exist without expending and consuming energy. Therefore, we need to view things as an ecosystem. How can we think about an ecosystem in terms of its bio-elemental energies, which are participating in the planet? The era of the Anthropocene raises the question of the terrestrial city. This city is an inhabitant of the Earth. This stimulates us to think about new ways to "create projects" because, given the current level of complexity, we really are going to have to imagine terrestrial cities, atmospheric cities... The challenge will be both political and about the living ecosystem. Not just one or the other.

THE TSUNAMI OF DIGITAL TECHNOLOGY

AM – A tidal wave is threatening the city. It is digital technology and artificial intelligence (AI). In a few short years, these fields have become incredibly important to how we plan the city of tomorrow, although they have not been able to identify all the potential benefits or the full extent of the dangers. Innovation lies not in digital technology or AI, but rather in how they are used and applied. The power of the economic behemoths being constructed behind these technological advances is already immense. Meanwhile, digital technology is also enabling inhabitants to take back power. It enables them to share information, to cooperate, to exchange goods and services, to take action. It's important, therefore, to separate out all the different aspects of this economic revolution. Who owns these technologies that enable vast amounts of data to be processed? Who sells this data? Who uses it? Essentially, the Gafam companies (Google, Apple, Facebook, Amazon, Microsoft) have the power to bring about almost irreversible progress and innovation on all continents, a power that enables them to take control of our cities. This power is potentially very violent. All forms of violence provoke reactions like those we are currently seeing in regard to the protection of personal data. Things might have gone differently if some of this personal data could have been transferred to a collective – the city – and its use restricted for the benefit of this collective; this could have led to some truly innovative communal initiatives. This implies a different kind of system in which these new technologies would facilitate the trading of services. If this service-city retained control of this data, the benefits for the inhabitants would be immediate and would transform their way of life. In this kind of horizontal system, inhabitants could be both producers and consumers of services. This is not some kind of utopia – it's happening right now and it's a profound transformation.

At the moment, in terms of the architecture of buildings, when shared space is proposed it is in response to a specific demand and does not curtail (on the contrary) the profitability of the property investment. Thanks to the city's inhabitants, we will (re)discover the importance of the "commons". Now that we have all these new ways of being proactive in the city, we need to make sure these opportunities are not distorted and corrupted by bad decisions regarding the control of their exploitation.

CY– The image of the tidal wave illustrates how an extraordinary power can be destructive. The new possibilities available to humans are a double-edged sword – they could have the best and the worst of outcomes. This is the notorious Pharmakon, first introduced by Plato and revisited by Derrida and other authors, notably Bernard Stiegler. Stiegler analyses and validates the idea that new powers could enslave us, but could also be our liberators. It's either collective and/or individual empowerment, or collective and/or individual subjugation.

In making machines capable of intelligence and even feelings (the possibility is being explored) we are playing God. We cannot guess at the medium-term and long-term impact this will have on the processing, generation and transmission of data. These paradoxes in which we find ourselves fall under the scope of both top-down and bottom-up activities. With the Gafam companies, just a few individuals control the information. However, and this can be seen in the fields of architecture and urbanism, the transformations brought about by digital technology also regenerate opportunities, inspire possibilities, reinvent practices like bartering, and revive tools for sharing and solidarity between individuals who know how to react quickly with this digital culture and artificial intelligence. If you view the city from this perspective, we are indeed being carried off by an unstoppable tsunami. We are, in fact, "inside" this digital culture. Those who cannot speak the language, who can't use it, will be expelled from the system. This brings us back to politics. In *Le Contrat social* [The Social Contract], Rousseau, one of the forerunners of the French Revolution, wrote: "Man is born free and everywhere he is in chains". It's an illuminating phrase about the fight for justice. Digital technology reignites questions about the nature of municipal justice, environmental justice, social justice. In the Age of Enlightenment, justice, science, education and politics were associated with the hope for a universal that could be shared. In a way, these digital technologies are (re)introducing the ability to share information worldwide, almost instantly.

THE INTENSIFIED INTELLIGENCE OF THE CITY

AM– We need to examine the political impact of the digital revolution. Will it result in a new democracy? That is to say, a territorial democracy of the city? Or will democracy wither under the yoke of a "technical dictatorship of social organization" driven by large multinationals?

First, I would like to examine the possibilities AI (artificial intelligence) could have for the city. What smart solutions we could use to organize our lives together. Fundamentally, what we all like about the city is its urbanity, its intensity. We are happy when the city is energetic, we like its hustle and bustle. The terrace of a lively cafe, a busy street, a thriving market: they all produce the joy of the encounter, an interpersonal energy. And people like these types of urban energy. So, what if we created more of it? What if we made it more instinctive? What if trading services with one another became so normal that it was no longer noteworthy? Of course, what I'm describing is a utopia. But I want to underscore the idea that digital technology could bring about a utopia of new forms of mutual service. Every individual would find their usefulness in society, everybody would contribute to a better society for all.

Let's return to the political impacts of digital technology. It is, of course, possible to reinvent democracy via direct democracy, i.e. democracy that operates on a single level. Is this any worse than having intermediate representative bodies? The question has been asked. In universal suffrage, each person's point of view is respected, it is equal to the viewpoints of others. But you have to go through the entire system. This doesn't mean that there aren't alternative ways to explain things, to bring about changes in opinion with better educated, more informed citizens. There is a case to be made for a new political utopia; it is within our grasp, even if it has not yet been achieved. I predict a renaissance of political democracy in which everyone's views count. We are all capable of managing everyday facilities. That's why I support the service-city. What we need to create is a kind of unforced cooperative.

CY– Effectively, we are not just biological organisms, we are also societal and political creatures, and we have to decide what type of society we want. And we are at a crossroads. How do we want to live together? How are we going to use and regulate data? Anything that can move us towards collective intelligence is beneficial for understanding and sharing. But the ability to make decisions together should not be reduced to abstractions. It must take emotions and feelings seriously, because they form part of our humanity and our psychophysiology as empathetic, social, emotional and cognitive beings. We do, of course, need to resist emotional manipulation; we need to reconnect with our ability to collectively manage information that can be integrated, exchanged and transmitted. Because this is also part of what makes a city. And it is a heuristic that Europan teams grab this ability to make projects.

AM– When you organize data, you produce a kind of intelligence. From a shapeless mass emerges information that helps us to react, to make decisions, and sometimes to change our behaviour. Digital technology gives us a way of providing for ourselves. It's easy to get lost in the vast amounts of information, but it's also possible to control it and turn it into something beneficial. Meanwhile, AI (artificial intelligence) allows us to leverage the quantity and diversity of this data to produce a huge range of outputs. And who is responsible for making the most of this data? Us – the public authorities together with its citizens – or those to whom we have relinquished our economic power?

CY– This great wealth of intelligence, in the sense of processing large amounts of data, could prove dangerous if it is not accompanied by the ethical and aesthetic dimensions of society. This is, incidentally, the essence of habitable and equitable development. In other words, it's us, humans, who decide to become a new entity, to share our intelligence, power and weaknesses. And that is a political choice. Otherwise, this could very quickly lead to two completely disconnected universes: a world of slaves, and a world of so-called experts who will own and control these new opportunities. We are at a crossroads.

AM– As I'm listening to you, I'm thinking about the possibility of Gafam seizing the power of digital technology from the city. I think if people came together, at the level of buildings, neighbourhoods, cities, we could build a resistance movement to recapture this power – a power we haven't fully lost yet, and which we should protect, as it will be a source of our strength. That's why I think it's better for the city to embrace this tidal wave rather than try to fight it or hold it back, because that would put us in a really weak position.

CY– It would be madness to think we could ignore this digital revolution, because humanity never goes backwards when it comes to its inventions, it always uses them to go ahead. But, this empowerment, these new possibilities, call on us to work together to ensure that we don't become slaves to the economic power of private interests.

This brings me back to Darwin. There aren't many publications that discuss the importance of cooperation in coevolution, but he addressed it in his works alongside competition. Cooperation is a resource. It can be found in the world of living things but also in technology. Technological systems created by humans also co-evolve. They inform one another, transform one another. Robots have this capacity to learn and transform. But how do we bring about restorative synergies that benefit humanity as a whole? In other words, in a way that does not allow the law of the survival of the fittest to prevail? That's the perfect utopia. But we can get there together; because we are together. In *The Plague* by Albert Camus, the heroes are those who risk their lives to help others. A doctor treats patients while he experiences aching loneliness, but he makes no grandiose speeches. In the face of this plague, which could destroy everything, there are people who keep on trying to help those in need. In both of these strata, living organisms and technology (including the most advanced technology, like AI), are collective tools. They could be usurped by others, just as wealth has been. This ability to create together ultimately becomes a political act that should be encouraged.

A PERIOD OF RE-NAISSANCE FOR OUR CITIES

AM– With the arrival of big global questions about society, such as sustainable development, COP21 and the greenhouse effect, cities occasionally stand in for States. Faced with these global issues, cities, rather than being considered subsections of States, have drawn a certain power from coming together, as they did during COP21. Now that civilization is primarily urban, isn't it time for a rebalancing of power between Cities and States?

How does this development relate to the Renaissance? During the Renaissance, transformations took hold first in the cities, or rather city-states. The transformation we are seeing today, driven by decentralized renewable energies and digital technologies, seems to me, as during the Renaissance, to be better suited to cities than States. In my opinion, that's good news. I call it re-birth (re-naissance) but it could also be called "fundamental transformation". Cities can find new balances and alchemies of transformation more easily than States. There is more permanence in cities; democratization is more direct and more natural. With this idea of re-naissance, I am alluding to the power of Cities to resist States.

Incidentally, during the Bâtiment Responsable (RBR 2020-2050) working group on responsible construction, our discussions with the architects – I'm thinking about Anne Demians in particular – led us to the same hypothesis about a period of revival. In terms of Construction, there are very powerful forces of transformation at work. A vast range of new materials and components have been developed; innovative design and construction techniques have emerged (the use of virtual models, robots, exoskeletons and 3D printers). At City level, meanwhile, the technical innovations are innumerable and very promising; they relate to energy, networks, transport, the circular economy and logistics. How do we transform these into a cultural project, a political project of the city? Once again, the scale of the city lends itself better to extensive transformation than that of the State. That's why the City will be at the epicentre of this re-naissance, because it has all the ingredients required for regeneration.

CY– As we've already discussed, the city is taking up more and more space on this planet: spatially and economically, but also because it has reached huge proportions in terms of its sprawl and huge size. But in a more general sense, beyond saying that these entities we call "cities" are going to disappear, what can we say? The urban certainly is taking over, but cities are enduring and still have a strong power. There are even possibilities that there will be a revival of the city because embedded in its DNA is the idea of the human settlement that regenerates from within: a group of humans come together and agree to create a human settlement. This happens in villages and in the countryside too, but cities, particularly in the West, carry the notion of citizenship. European and American cities have been hugely impacted by the citizen-democracy relationship. How can we stimulate democracy?

The re-birth of cities requires the political ability to situate the challenges we face within spatial justice. In his book *Le Droit à la ville* [The Right to the City], Henri Lefebvre highlighted that which was at stake in respect of this right. He discussed relationships of citizenship and citizen emancipation. Now, Edward Soja's publications on the topic of spatial justice are very stimulating. The idea that the city is one of man's greatest inventions because in it he has found a political way of coexisting, of developing this cultural dimension derived from multi-otherness. Although the re-birth of cities may touch on this idea of the re-birth of spatial justice, it also demonstrates that, in the face of problems, the city is obliged to invent and to start over. There is a re-birth because the city starts over as a city. And this new beginning is articulated around values like fairness, which is a challenge for property development projects.

Cities have always welcomed great diversity. If they start to turn inwards (or towards certain categories of privileged inhabitants), they no longer conform to the idea of the city, the spirit of the city. Just as we speak of genius loci, there is also the spirit of cities, which incorporates a desire to live together and to invent democratic political tools. In the philosophy of Greece in the 5th century BC, the Greek city was designed to be the right scale to enable coexistence, co-leadership and coproduction. There was a strong relationship with the scale of the territory.

What about the scales of the cities? This 21st century re-birth or re-naissance requires the city to remain outward-looking and to face up to the reality of its scale. It's a challenge. Throughout the world, we can see the re-birth of communities. These are not closed-off communities but rather scales that are more conducive to assembly, to governance, but also to shared everyday life. There is, here, the potential for a re-birth of urbanity and citizenship.

AM – The re-birth of cities could also be considered a reaction to an individualistic and consumerist society, in which happiness is satisfying our appetites for ownership and, via aggressive marketing strategies, creating needs we don't actually have. Personal happiness becomes the basis for social questions. However, one of the best things about the city is the impossibility of being totally alone. There are people everywhere, a density, an effective otherness. What can you achieve with other people? What can you change together? How can you draw strength from this human presence? It's also in this sense that I speak about re-birth. It's the possibility of rediscovering simple pleasures, opportunities for exchange and sharing outside of economic mechanisms. This could be neighbourhood parties, rediscovered commons, or any kind of public space created by residents. Some would call this naïveté. I don't agree. Rather than owning everything individually, people share a certain number of things. And the city is the sphere in which this sharing takes place. It enhances the value of proximity, the benefit of living alongside one another. This proximity provides a level of fulfillment that trumps the dynamic of personal egoism. It's a kind of laboratory for "making together".

MAKING TOGETHER

CY – It is striking to see how some movements reappear within individualistic societies to reaffirm the belief that solidarity is the best way of overcoming difficulties. We are currently seeing a revival of forms of local solidarity, collective willpower and shared desire. The notions of justice and the communal are also becoming more closely linked. The idea of fairness, which was revived thanks to the sustainable development movement, is also the desire for a shared life that forms part of a chain of intergenerational transfer. A kind of libido *communicandis*. Several publications and Europan projects focus on this notion of cooperation, of collaboration, and of making together. It applies to the digital culture we've discussed here but also to other fields such as education, business and healthcare. The desire to find common ground opens up new avenues for sharing.

AM – We should remember that the word "Renaissance" was first used by French historian Jules Michelet in the 19th century when writing about the 16th century. It was taken up again a few years later by the art historian Jacob Burckhardt in his book *The Civilization of the Renaissance in Italy*. It is, therefore, an assessment, an idea constructed after the architectural, urban and artistic production of the 16th century. In the word "Renaissance" I see a kind of "harmony", a maturity, a fulfillment.

You spoke about justice, democracy, respect. The assimilation of technology by culture. During the Renaissance, scholars were at the same time philosophers and engineers and architects; today, the sheer volume of knowledge requires us to specialize. However, there's nothing stopping us from creating groups that enable people to work together to achieve the overarching vision of Renaissance Man. Now, in the 21st century, we can detect the harmony in the architecture of re-naissance, as in all the artistic production from the period. The idea of the re-naissance City could lead to new forms of architecture, street furniture and urban design that would symbolize this desire to live together in a different way. It would be an exhilarating challenge for architects and urbanists, given that morphology and metabolism are of equal importance.

CY – The Renaissance was characterized by this idea of looking for a new harmony through a kind of neoclassicism, searching for a new balance between things. The re-naissance or re-birth that we are talking about here obviously has nothing to do with that of the 16th century. If there is a re-birth, it's because we are coming to the end of an extremely invasive model, one that destroys ecosystems and social equilibrium, and because we are seeing the emergence of new perspectives that incorporate techno-scientific culture, education and a focus on humanity as a whole. It's also a reaction to the inequitable city: a phenomenon that has arrived, that is persisting and that is set to get even worse. So, how can we (re) discover a harmony that allows humans and non-humans to live together? If there is a re-birth, it will be because the current model is broken and unpopular, even though we've yet to fully define the new ways of making and acting. The city, birthplace of urban civilization, must now respond by transforming itself, adapting to humanity's new global challenges, and rediscovering its role as a life-giving force. The importance afforded to ecoagriculture, in particular the success of market gardens and the popularity of community gardens, is reinventing how we think and act.

TOMORROW: THE PROJECT, THE ACTION

THE ACTION PROJECT

AM – One of the central innovations Europan has introduced and developed is the idea of the action-project. The project is not a "fixed", fully formed project: it is a project of transformation. I'd like to focus on two key elements that I think illustrate the potential of the action-project.

The first is the temporal dimension. Urbanism and architecture have, for a long time, thought of themselves in terms of three-dimensional space only. I think, in the future, designers will plan in three-dimensional space plus the temporal dimension, i.e., in four dimensions, xyz plus t. It's not a question of applying Einsteinian physics, because the speed of the transformation is very slow and there is no space-time "curve"; however, time does interact with xyz.

The second is the overlapping of scales in the action-project – we need to stop talking about separate scales – , which enables a logic to unfold. On the site, we must be able to comprehend the scale of the building and the neighbourhood, as well as that of the city, to understand the dialogue between the city and its territories.

The action-project, using the variable of "time" as a conceptual input, puts forward the hypothesis that, in the future, action will adapt over time. I don't actually believe in the idea of an action planned in advance, because, as urbanism projects can take fifteen years (maybe more) to go from idea to implementation, the project needs to be able to adapt, and for this to be taken into account from the very beginning. Our comparison with living organisms comes to the same conclusion. The city, architecture, and the life of the building, are undergoing constant transformations. Buildings and houses are modified by means of constant additions that are influenced by the people that use them and how they live. This is especially true for the city. And these transformations are, without a doubt, happening faster and faster.

The action-project unfolds in all phases. It speaks of the site, the formulation of the brief, the spirit of the project, its implementation, the observation of its initial achievements, and the uncompromising criticism that may or may not change the original project, transform it, adapt it, or incorporate new economic opportunities and new factors that were ignored or did not exist when the brief was drawn up. It is the adaptable city of Europan sessions 11 and 12. This adaptation is ongoing; it goes beyond the tangible buildings. The action-project is founded on strategies, on the notion of maturation over time.

Europan does not have the monopoly on this approach but we have, I think, championed it more than others. I believe that our way of thinking, how we act and the way we engage the persona of the "architect and urbanist" has influenced the action-project. Teams that take on the action-project are in it for the long haul. They need to be on site for months, years even, to analyse their action in semi-real time. The site is like a patient that the urbanist architect constantly checks on to see how it is reacting. Is the patient well, does he have good mental health, is he happy, productive? We must dare to tackle questions of this kind. When a city is confident, when it has a positive image of itself, it has good mental health. Some cities are depressed. We just need to know how to spot it.

CO-EVOLVING VIA THE ACTION-PROJECT

CY– The changes we are seeing in the Europan projects underscore the impact of the action-project and the synergies with the becoming, a concept that took hold in the 21st century but was introduced and explored in depth by philosophy in the 19th century. I think that's one of the reasons Deleuze and Guattari[1] had such an influence on the world of architecture (and not just in Europe). They believed that things were becoming in such an active way that it was impossible for us to control and manage everything. Adding the dimension of time to the project is a step in the right direction. In his book *The Urban Village* and through his involvement in the territorialist movement, with its focus on bioregions, Alberto Magnaghi expounds the need to think in terms of coevolution. This produces highly complex ecosystems that require us to develop ways of understanding them using a wide range of data. Faced with equations where *x* is unknown, we will need even more data. However, new technologies allow us to store and cross-reference this data and, therefore, to leverage its power.

In terms of the potential future of human settlements, there are cities but there is also everything that is not city. This rural and urban eco-territorial issue requires us to invent new scenarios. One possibility is to focus on catastrophic scenarios, "we can't do anything, we can't control anything", which is a kind of passive acceptance of

1 Such is the simultaneity of a future that its central function is to escape the present. As it escapes the present, the future "cannot abide the separation or distinction between the before and the after, the past and the future"; Gilles Deleuze and Félix Guattari, *Mille Plateaux*, Paris, Éditions de Minuit, 1980.

the end of humanity, like the catastrophic scenarios envisaged by Paul Virilio or Jared Diamond in his book *Collapse* [2004]. But in architecture, a key possibility is to know how to produce scenarios of birth and re-birth as regeneration. Scenarios of regeneration are active, they propose actions to make together, to connect differently in ways that are more complex, more harmonious, and better suited to the challenges we face – such as ways to remake the city and the countryside, to farm, to feed ourselves, or ways of dealing with climate change.

AM– The word coevolution is interesting. In physics, objects move and their movements influence the movement of other objects. They interact and co-evolve. How can we act on the city using an action-project? I could set things in motion and allow the co-evolutions to materialize. I just let them happen, I don't stop anything. I drop a stone into a pond and let the ripples spread out. There's no question that I have created something. The ripples reach the edge of the pond and flow back. That's the first idea. In the second, I drop a stone and ripples appear on the surface of the pond, but then I drop a second, and new ripples appear around this second circle. When the ripples from the first circle meet those of the second, they interact. Then a third stone is dropped, etc. I become a programmer of coevolution. It's a way of giving meaning and legibility to the word "action" as it relates to the project. The project triggers the process of co-evolution and, gradually, in response to action, the various co-evolutions are modified. But we need to know how to observe these co-evolutions. When you throw a stone into the water, the ripples spread far out. They don't just stay in the immediate area. It's the same with an architecture project. If the ripples reach faraway shores, I should go there to see what my action has produced. The idea of coevolution reaches its full potential when it is not limited to a small area.

HARMONIES

CY– Rethinking coevolution is also a way of rethinking harmony in terms of synergies. The philosopher Daniel Payot describes the current state of the world as "post-harmony"[2], and I agree with him. However, that doesn't mean that there are no new harmonies to be invented. On the contrary, our aim is to find out what kinds of harmony are made possible by taking into account the element of time and an understanding of the whole and the individual. What landmarks can we use to navigate through the action? Because, in essence, the project is a beginning, a starting point, a hypothesis of transformation. In Europan, this becomes a shared way of working, and then a shared action. Every action triggers social, physical and biochemical phenomena that cannot be ignored. How can we rediscover the harmonies between the forces of nature, life and societies?

AM– We need to stress that we are not advocating simply leaving coevolution to the "forces" of chance; rather we want to master this coevolution to tend towards greater harmony. In the future, it would be helpful to encourage the observation of the coevolutions caused by the action-project. Too often we need to wait for dysfunction to arise before we can look at ways to resolve it. Going forward, we will need to prioritize prevention. To do this, we will need to pay close attention to the progress of this evolution, this co-evolution of the city. And, therefore, we will need to develop tools for observation. This is an area of urbanism that is ripe for development. Just as we carry out blood tests to understand what is behind a change in a person's health, we need to invent the equivalent for the city.

2 Daniel Payot, *Après l'harmonie*, Belval, Circé, 2000.

CY– John Dewey, the pragmatist philosopher, underscored the importance of observation, of inquiry. This is always an aspect of good projects. The key to developing coevolution scenarios that are invigorating (and not malignant) is to create instruments of observation and description as well as instruments of measurement. For a long time, biodiversity was not a matter for observation. Now it is a central focus. In these potential scenarios, it is important to identify the determining factors for both humans and non-humans.

AM– If you only observe three factors, for example, that means that the action will only apply to these three factors. If you are tackling the city in all its complexity, you need to act on many more parameters. You alluded to the storage of data. It's clear that the parameters we have taken into account in the past have, for a long time, been extremely limited by our inability to process data in large quantities. Now, this constraint is gone. We need to develop smart ways to enable us to comprehend the city in all its complexity, in its entirety.

CY– Absolutely. We will know how to invent the right tools when we have defined our methods and objectives; when we know where we want to go. Observation can sometimes be a lesson in humility. We have to acknowledge that reality is a complex becoming. We need to become familiar with reality, to learn how to recognize it in its own dynamic. Places are multiple entities that are permanently changing. We need to understand and recognize that which is already there, the existing. Except that the existing-already-there is shifting. We need tools of representation and imagination: observations, descriptions, inquiries, fictions, etc. in order to comprehend reality in all its spatio-temporal breadth and not as it has been proclaimed to be. Space, the "spatial turning point" as Michel Lussault puts it, is the awareness of the limits of the Earth and the atmosphere and, therefore, the limits of terrestrial organisms. However, space and time are linked: it's not one or the other.

AM – The city is so complex; multiplying and interweaving various points of view enriches our understanding of human beings and represents a great leap forward in terms of the project. The action-project needs this constant enrichment from the here-and-now, from the observation of what is happening and what is likely to happen. It's one of Europan's unique features. It forms part of "its genes". The articulation of scales is a key requirement of an action-project. The effect of the project must be examined at all levels. To increase its impact, we need to identify the actions that will occur concurrently in its sphere of influence. The project is composed of what happens on the proposed site and all the actions associated with the resulting effects. And *vice versa*. The action-project may be focused on a clearly located site, but there is no limit, no territorial boundary. It incorporates all flows. In addition, there are many who now consider the city to be simultaneously composed of flows and forms. This demonstrates the need to also consider the conduits that these flows pass through.

IMMERSING OURSELVES TO SEE, UNDERSTAND, FEEL AND EMPATHIZE WITH

CY – Flows are also forces. Zygmunt Bauman, in examining what he calls the "liquid society" or "liquid modernity" (as opposed to "solid societies"), highlights the adverse effects of consumerism, and the atomization and manufacturing of fears and insecurities. If we use the Europan projects as a springboard for a prospective analysis, it seems that three key areas of focus or fundamental challenges can be drawn from the research and practice. The first has to do with the idea of immersion. In terms of the action-project, Gilles Clément explained it perfectly when he referred to the principle of immersion as primordial. We are not situated beyond reality; we are in it, we form part of it. An animal's power is its immersion. A cat has the capacity to find the place best suited to it. It is completely immersed; it is bound to other things. Humans designed cities and other settlements using skills and techniques they invented, but they also often placed too much importance on an objectifying distance. A contemporary re-naissance will be achieved by means of immersive processes that re-situate us within the scales of life and the universe that incorporate the everyday and a changing world.

A second key area of focus is the linking of forces to one another. How do we bind things together? We use principles that are not necessarily from the world of construction and the material. We rely heavily on psychological and spiritual aspects. Binding multiple elements together is an intrinsic part of any architecture that chooses not to start from scratch. Co-evolution can be disruptive, but it forms part of a process of regrowth and repetition, or, as Soren Kierkegaard calls it: "remembering ahead. Remembering the future in order to exist in the sense of becoming. We make use of that which we inherit.

A third avenue of exploration relates to empathy as the ability to share the feeling of other people but also those of other forms of life. It is, therefore, another expression of interlinking. Connecting, coming together as part of a project, is both work and action. Work, which Hannah Arendt differentiates from action, has the power to cross time and can unite us as humans. It is works that are passed down through the generations, because they are capable of enduring. In an encounter with a place, there is both a contextual and an idea-driven dimension.

We need to make choices in order to live together: the principle that it is not the law of the survival of the fittest that should prevail but rather the idea of the pact and, therefore, cooperation, solidarity, kinship. These various arguments relate to the world of values. Politics relates to choices. Rejecting slavery, or child labour, these are values. Worrying about future generations is another.

AM– What that means is that if we are involved in an immersion-project, the person in charge of implementing it should empathize with the others involved. In Promethean thinking, there is no desire to be with others except for the desire to impose a task using force. But once you are immersed, even the way in which the project is managed must be respectful of the project itself.

CY– And this leads us to consider the various correlated facets of our relationships with environments. Today neuroscience is studying how the human brain creates synergies between the socio-affective-cognitive dynamics. Much work has been done on the cognitive dimension but we shouldn't forget the emotional aspect. It has been demonstrated that emotions impact whether or not the cognitive dimension is stimulated. Our emotions clearly have significant power.

AM– The way in which action-projects are carried out should be consistent with the content of the action-project itself. How we get people to participate, that's the key. This proactive participation makes inhabitants co-authors of the transformation and co-actors in the action. We need to immerse ourselves to gain a full understanding and, of course, to propose the best solutions. The action will be discussed, debated, transformed, adapted and the project itself will evolve through contact with others and will be adjusted as a result. The action-project requires immersion, empathy and cooperation.

TOMORROW: THE PROFESSION(S)

A BIG BANG FOR PROFESSIONS

AM– The professions are changing because the city changes and mutates; the action on it changes, and the action-project and its new logic modify the role of the designer. We also need to bear in mind that economic stakeholders are seeing their professions undergo a metamorphosis too. Essentially, the architect-urbanist-landscape architect used to just have one interlocutor, their client, the City. Now, the relationship between private and public financing is having an ever greater influence on the feasibility of projects, a topic that Europan has helped open to debate. And private funding means private stakeholders involved in the development of city projects. What is being decided, then, in the discussions, is the amount of room for maneuver that public power should maintain. Private stakeholders may want to put forward their own vision. The designer needs to listen all parties, and his job is to understand their aspirations and to translate them by means of the project. Session E14, Productive Cities, also posed this question. Rather than discussing the future of all of professions involved, we are going to focus in particular on the role of the architect-urbanist-landscape architect.

CY– For a long time, the role of the architect was associated with that of an orchestra conductor. The architect was the master figure, the cornerstone of a group of players. Now, the professions of architecture have been pluralized and this one term groups together a number of hybrid professions in order to adapt to changing contexts, their complexity and the diversity of stakeholders. Today, architects train to be architect-urbanists and landscape architects, and schools of architecture provide this pluri-education. Europan has worked hard to ensure that these universes remain connected to one another, thanks to the interweaving of different scales. After all, designers are more hybridized than orchestral conductors: they are experts in more fields, and their role requires more collaboration. Moreover, hybridization is a prerequisite to working with interwoven scales. It seems to me that these approaches are becoming more and more prevalent. And that's a good thing for projects.

Faced with the complexity of a subject like "the city", designers now often expand their initial training by taking professional development courses at institutions like Sciences Po, in Paris. In professional terms, some of these courses focus less on the implementation of projects and more on mediation, recalling the role discussed by Deleuze and Guattari, the intercessor. Intercessors are intermediaries between different fields and forces. They establish points of interaction between different worlds. They are able to propose a translation, an interpretation that enables this interaction to be productive. This differs from the role of a mediator in terms of action. These translators between one world and another are like intermediaries between professions and groups of people. In my view, they will become indispensable.

AM– Are architects-urbanists-landscape architects working, or will they be working, in the service of the City and action-projects, as defenders of the public interest and guarantors of the quality of the development? Or will the city be, as it were, the territory that frames the work of the architect? For us here at Europan, the challenges of society, the evolution of the city, its transformation, its setting into motion, the goal of making it happy and successful all take precedence over "creating a work". This is important because the sites proposed in the context of Europan are often difficult and inhabited by vulnerable populations, sites on which the various actors must build with humility and responsibility. What is the role of the designer in these situations? He or she is just one individual among the many partners, inhabitants, associations, etc. I completely agree with you about the need for new ways of being. It's an important question. Architects-urbanists-landscape architects find themselves at the heart of a societal project in which the city takes a dominant role and imposes its personality. The role of demiurge, of lone conductor needs to be abandoned.

As soon as you start working in service of the transformation of a city, you need to assemble a collection of increasingly hybridized skills. Without being omniscient, you need to be culturally aware of all the economic, social and technical dimensions and how they fit together. You need to create synergies from the world of the economy, culture, representation, imagery; to prioritize empathy, living together, and finding forms of harmony. The designer needs to be able to measure, at least intellectually and culturally, all the dimensions of the city.

In the action-project, the action on the city inflects, accelerates, modifies or preserves. This presupposes the capacity for observation. If we go back to the comparison with a blood test, who analyses the results? The architect-urbanist-landscape architect should be able to.

We've described the city as a living organism that wants to be in good health. It may be become infected with disease, including morphological types of diseases. Bone diseases need to be treated, fractures mended, and limbs and skeletal structures (re)created, etc. This is the surgery of morphology. But there are also the surgeries of metabolism. Soft tissue surgeons treat the liver, the stomach, the kidneys, and the internal functions, including the heart and body fluids. And, as in medicine, there are those who treat patients using acupuncture to work on the nervous system. In Europan, it's common to see action-projects that make explicit reference to methods of acupuncture. This vocabulary, these metaphors borrowed from medicine, make sense. They illustrate the breadth of knowledge of these architects-urbanists-landscape architects, who need to know about morphology and metabolism and, at the same time, to take an interest in the brain and the mental health of the city – a living being with its own personality. It's maybe going a bit far to talk about the brain and mental health of a city. But AI (artificial intelligence) is starting to impact the city. We need to understand the implications.

CY– Given pollution and climate change, questions around health actually are central to the challenges for urban development. There is a growing focus, in all fields, on the concept of "taking care". From architects to company managers, from nurses to teachers, this means having concern for others, concern for the world. The scales of the city and, more broadly speaking, those of the territory and the universe, have profoundly transformed the architectural professions. When Bernard Reichen called himself a territorial architect, he conveyed an alternative vision of the profession: acting in service of the territories by working with a multitude of stakeholders. The ability to see and listen – to listen not just to oneself but to others – is fundamental. To consider not just what you already know but to engage with otherness and even participate in collective creation.

The profession of architect will endure because it is capable of adapting. That's why the collectives that often form these days out of architecture schools bring together a range of different individuals, to better enable them to manage projects. Georges Bataille was harshly critical of architects that only worked, as he put it, for those who have the power. Architects are at a turning point right now; they are looking to develop methods that involve the stakeholders of inhabited environments, and to immerse themselves in situations, to communicate and intercede. Because architects must also be intercessors and mediators, able to listen, to explain what they are doing, how they are doing it and why, and to explain what others are doing and what they expect. But they are not oracles even if part of the intercession is about listening to places and the Earth. It was Édouard Glissant who said "Act locally, think globally". And in order to act locally and think globally, we need to be intercessors for the world. So there is, necessarily, an aspect of being in tune with situations. Many collectives have aligned themselves with this movement. There were the pioneers in this area, such as Lucien Kroll, Patrick Bouchain and his teams, or Philippe Madec, who immersed themselves and sought out contact with people and situations. Their project strategies and creations are nothing alike, but they all choose to position the architect not above other people, but rather with and among them: with the decision-makers, the inhabitants and the other stakeholders. Architects can incorporate into their projects elements that are fragile and intangible but very precious. They can communicate the value of certain decisions, for example, explaining to inhabitants the reason behind the details, the forms, the precise location of a window which guides the sight outside... Inhabitants can also feed back their expectations, they are empowered to say "yes" to a project, even to contribute to it. Not every idea has to come from the client or the designer. It's important for the designer to be humble and modest, but also to take their pride in their work. This is a kind of sharing that, in order to be achieved, must use specific tools to enable the various experts to find "the skills of the inhabitants" as Michel de Certeau puts it. This creates shared projects. It's a way of taking care: care of the habitat, social or otherwise.

A COLLABORATIVE APPROACH

AM – We are going to see a change in the methods used in architecture and urbanism. It's the shift from a sequential approach to a collaborative approach. It's a major development. It is made possible by the construction of a virtual model in the upstream phase of the project. Before any construction has taken place, you can study the consequences together with all the stakeholders. This technology has arrived in our fields, but it's still just a tool.

Let's remind ourselves of the sequential chain of events in the construction of a building. A client chooses their architects, then the design offices, then the contractors, who obtain the materials from the manufacturers. For a long time, this sequential order has been immutable although there has been a desire to improve the interfaces. In collaborative work, everyone retains their role. There is still a client, a project manager, a contractor, a manufacturer, but they interact within a collective project. And it's important to recognize that Europan is in currently imposing this vision in the urbanism field. A Europan project must follow this kind of approach. It begins with a collaborative dialogue between the client, the project manager and the Europan organizer, following which the client may decide to change their brief. When the client hears what the participants have to say, they may see that the question could be asked in a different way. The competitors' responses push them to reconsider and to develop. Collaboration is more effective, more productive and it possesses a potential intelligence that the sequential approach does not. In effect, this collaborative approach leaves space for the inhabitants destined to live in the future project. It is not simply consultation. We need to push for this approach of co-construction, which requires the designer to understand the wishes of the inhabitants before they propose an interpretation.

In the collaborative method, once each individual retains their profession, it's no longer an issue. The profession remains, but it is exercised differently, in another context. For the architect-urbanist-landscape architect, the way in which the project is expressed becomes fundamental to the communication of their intentions. The form in which the design is presented is key to making themselves understood. In the construction industry, the main language used to be the working drawings. However, in a collaborative approach to urbanism (although this also applies to the construction industry) the design phase is often enriched by input from the stakeholders gathered around the table. It is, therefore, necessary to use forms of expression that can be understood by everyone, not just professionals.

In the Europan responses, we are increasingly seeing projects depicted as transformation-scenarios and descriptions of ways of life. Action-projects presented in the form of screenplays. Long-term developments brought to life by human avatars. "This is the home of Mr and Mrs X; they have two children; Mr X works here, Mrs X works there, here's the nanny, this is how they do their shopping. This is how they live." Some candidates have even chosen not to design any new architecture, instead creating collages of existing buildings and proposing a "backdrop" – implying that the problem lies elsewhere. This led to heated debates between the members of the Europan 14 jury. For some, it was unthinkable to reward teams "that don't even know how to draw", while others defended "their refusal to settle on an architectural image" for a building or project that wouldn't take shape until many years later. These were interesting discussions about exactly the kinds of issues we are currently tackling. Personally, I'm not only in favour of the action-project. We need to remember that, at the end of day, there are buildings to be built, there is street furniture to be designed, there are cities to be created. That doesn't contradict what we have explored in relation to metabolism, flows, etc. But the city also exists through its morphology. The chosen mode of representation is a tool for communicating with others. Incidentally, we are currently discussing the possibility of asking entrants to submit videos, pieces of film that illustrate how people will live in the action-project of a future neighbourhood. And these moving images will facilitate dialogue with inhabitants. It's a new space for presentation within the collaborative approach. I think it creates a wonderful space for creativity.

The designer is, or should be, the person who creates harmony in the face of all the various contradictory economic and social forces and opposing desires and interests. This idea of harmony, in the sense of a utopia – a utopia of life in the city, of building a society in the city –, needs someone that can create harmony: this is the irreplaceable role of the architect-urbanist-landscape architect.

LIVING ON EARTH: A NEW ALLIANCE[3]

CY – Europan has been a pioneer in the use of collaborative and cooperative tools in the application of the concept of immersion at all stages of the process of co-design, co-implementation, and even co-evaluation. There are potential forms of collaboration in which the architect-intercessor will have to identify the challenges and even help evolve the brief at the start, and then potentially change it later on in the process. The winners of Europan often had to shift their client's expectations; they stayed in contact with sites and stakeholders; they figured out how to read between the lines and find a middle ground. They may also have to be intercessors at the end of the process, remaining in contact, explaining things properly and ensuring that the results are appropriable. I believe in the importance of creating synergies from this collective search for eco-harmonies; in the sense of bringing together very contradictory positions. This is not the harmony of fake pacification, of marketing rhetoric that gives the false impression that everything is perfectly calm. On the contrary, it involves working to reconcile contradictory points of view and interests. A project's quality depends on the ability of the designers to tackle conflict between stakeholders; to point out things that are easier to ignore; to make the most of situations. This cooperation revives the principles of desires and realities in all their contradictions: solving conflicts together, making living scenarios acceptable. In his book, Errance, Raymond Depardon speaks about the "acceptable place" as a way of making with.

3 Ilya Prigogine, Isabelle Stengers, *La Nouvelle Alliance* [1978], Paris, Gallimard, coll. "Folio/Essais", 1986.

In times like these, how can we "live in the present" of planet Earth? This question goes to the heart of how we think about new harmonies between nature and culture. This is our construction site. How can we find a synergetic harmony between the artefact and the forces of nature? There are so many avenues for exploration: palimpsests, climatic elements, energy transitions, digital transitions, zero carbon goals, biosourced materials, democratic re-births, etc. How do we invent in the midst of such diversity and plurality, and in a way that holds everything together, and by exercising our professions in multiple ways? By imagining the possibilities for coexistence and cohabitation between humans and non-humans. By writing these storylines. Today, in these times dominated by instability, the conflicts and catastrophes we face are becoming unbearable. We need to reinvent, imagine, dream of a world of possibilities in order to create, using our ability to reason, to discuss, to share, to feel, to dream. This dimension of architecture, in addition to making alternative viewpoints acceptable, is also a political, ethical and aesthetic dimension that has become crucial. In this Anthropocene Age, which was preceded by an era in which so much was not just invented but also destroyed, we need to examine what is acceptable. Europan is the perfect tool to address this challenge, together with young architects-urbanists-landscape architects who are ready to put forward hypotheses. There is always something conceptual, some little miracle in an architectural project. And there is also the contextual and the processual aspect. Europan is a pioneer in its exploration of the extent to which the conceptual, the contextual and the processual are inseparable. You could take it a step further: conceptual-contextual- processual but about the planet Earth version.

AM– The Planet Earth Version. This is the next course to set; and it requires us to rediscover the harmonies between the artefact – that which is built by man – and the forces of nature. What's new is that the dominant man on the planet is the man of urban civilization.

If we look at the history of Homo sapiens in terms of our species' relationship with Nature and the Earth, we can say, unequivocally, that we entered a new era at the end of the 20th century. Following the hunter-gatherer period (nostalgia for a lost Eden), the era of rural civilization (farmers, herdsmen) that had established a balanced relationship with Nature, in the sense of a nurturing and cultivated nature, faded away.

It was followed by the era of the "urban" man, a new urban world that replaced the rural world. At this moment in time, the urban man still needs to create a new alliance with Nature, and it's this central question that we have to solve.

• • •

In this discussion, we explored the future of City and Architecture by structuring it into a three-step process.

The first section, The City of Tomorrow, actually relates to the future of urban civilization, which must change drastically in order to cope with the incredible growth of the global population and to allow the Earth's inhabitants to live in harmony with the planet's resources.

The prognosis is grim; all scenarios involving gradual evolution are unacceptable; only rapid and radical change can be considered. We need to enter a new cycle, which has, incidentally, already begun.

In order to envision these changes, we need to start with the City in all its complexity, to examine all its economic, social, cultural and human dimensions.

This complexity is its richness, which is the richness of a "living organism". That is why the City will evolve like a living organism, passing on, over time, both its innate and acquired characteristics, which will act as subtle accelerators of transformation. This is such an immense transformation that we have ventured the hypothesis of a "re-birth" or re-naissance of the City, a new golden age which carries an implicit reference to the utopia of a regenerated City.

The complexity of the re-naissance City results in the complexity of the action. This is the second stage: action via the action-project, a process-project in three dimensions plus the dimension of time (four-dimensional space-time).

Our working methods, how we bring inhabitants together in new ways, how we enable new forms of democracy, will completely transform the life of the City. The City will become the testing ground for a new democracy, a symbol of the ideal political City.

These two utopian visions of "the regenerated City" and "the political City" entail a re-interrogation of the role of stakeholders, and in particular the role of architects and urbanists. This is the third stage.

These professions must draw on the complexity of the City and the complexity of action, they must acquire new skills, use new methods, find new balances. In short, they must be the architects of Harmony, building a new alliance between urban civilization (the urban world) and planet Earth.

"Déprise "[Dereliction], E13 runner-up for the Marne-la-Vallée site, Jonathan Cacchia, Cécile Frappat, Claire Girardeau and Louis Mejean.

Dereliction creates a fertile void in which the unpredictable can come and go as it pleases.

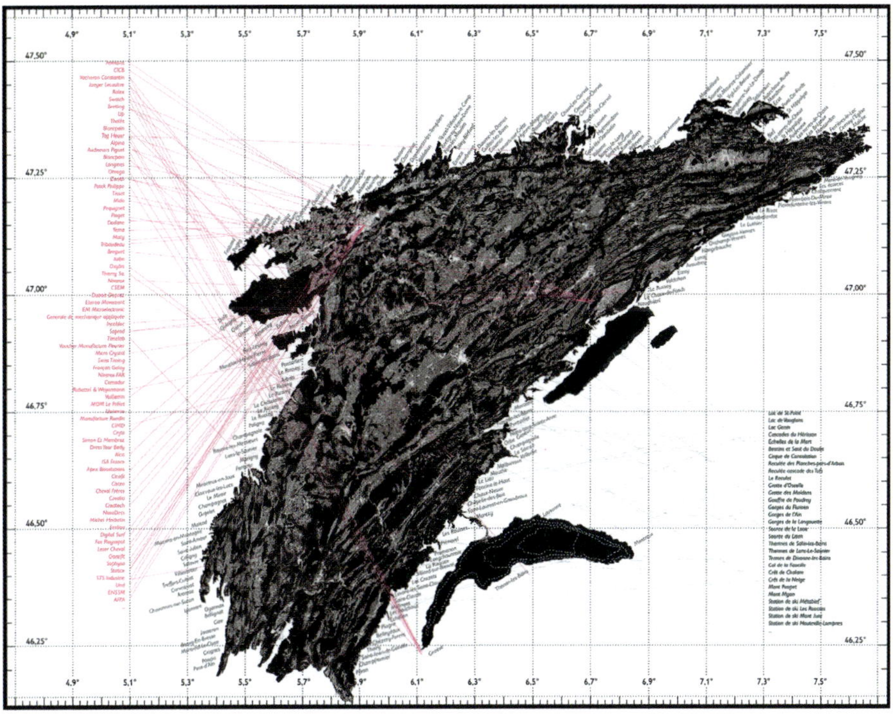

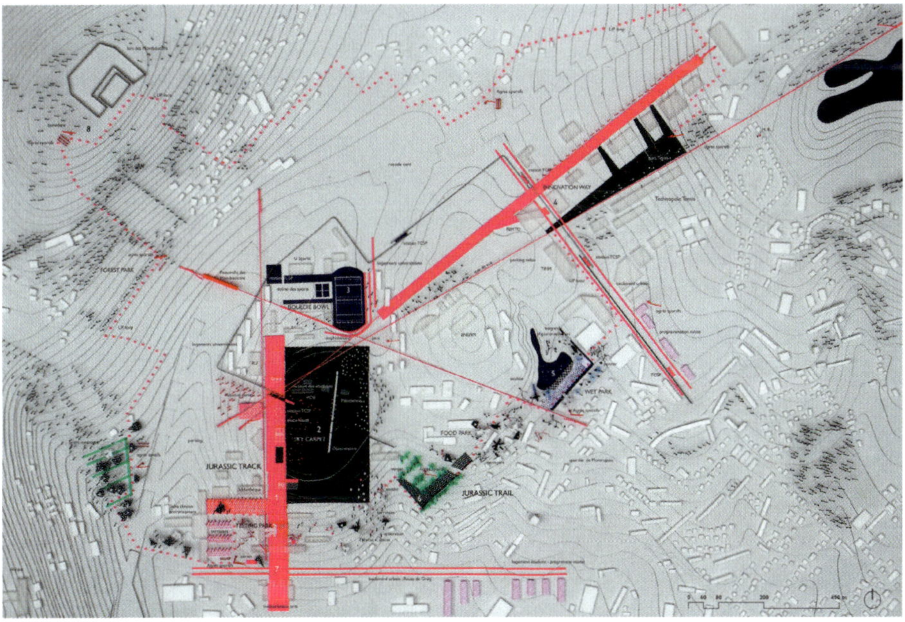

"Jurassic Parks", E14 winning project for the Besançon site, Benoit Barnoud and Clara Loukkal.

The project explored the relationship between a transnational economic strategy and the incidence of the Jurassic bedrock at three levels.

"Foam of production", E14 winning project for Amsterdam H-Buurt site, Timur Karimullin, Sofia Koutsenko and Timur Shabaev.

The project considered multiple levels of maintaining and expanding activity, and involved the community in an effort to maximize diversity.

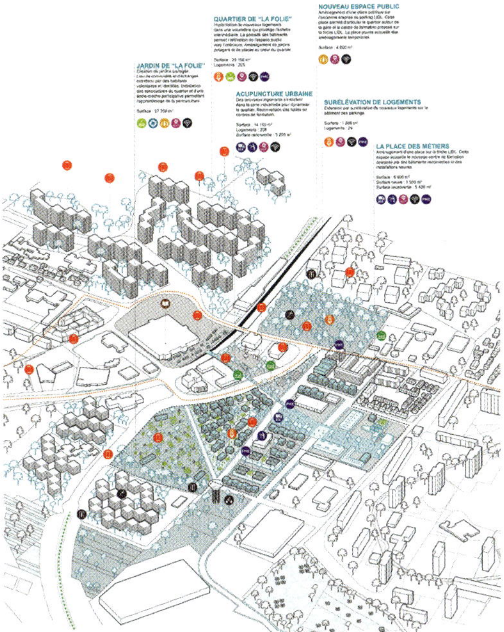

"Coopwork", E14 winning project for the Grigny/Ris-Orangis site, Diego Morales, Elaine Sanchez, Johana Salazar and Sebastian Morales Sotomayor.

This project enabled the emergence of new dynamics in which inhabitants, technology, and buildings interact. In this scenario, residents become the main stakeholders in the improvement of their environment, producing services for their own benefit.

"In motion", E13 winning project for the Barcelona site, Maria Langarita Sanchez and Víctor Navarro Ríos.

The management of each functional platform is independent and can be changed to accommodate future requirements.

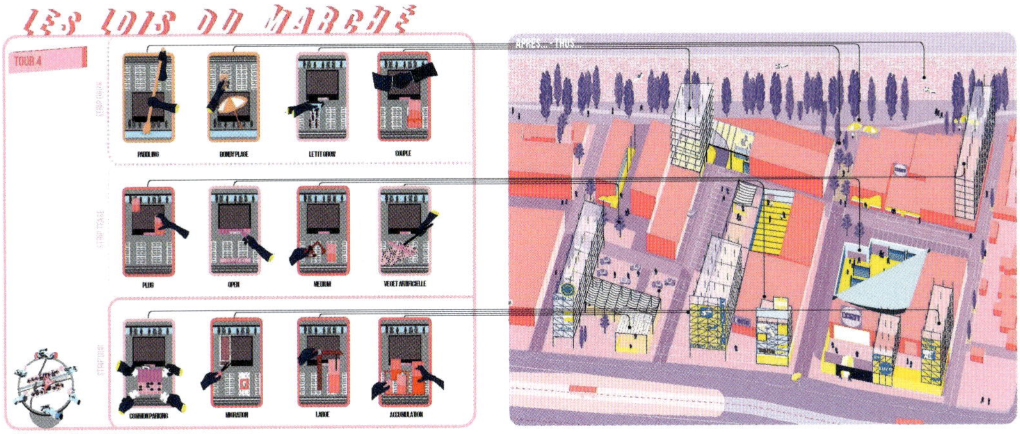

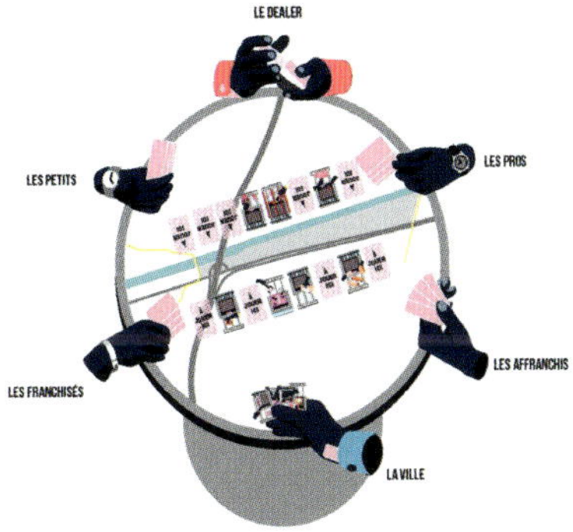

"Bondys' Count", E13 winning project for the Bondy site, Nicolas Barnavon, Marion Lacas, Jacques Ippoliti, Denis Brochard, Loïc Daniel, Quentin L'Hôte, Guillaume Barnavon and Charles Boucasse.

Participants come together on this new playing field without knowing in advance what is going to happen.

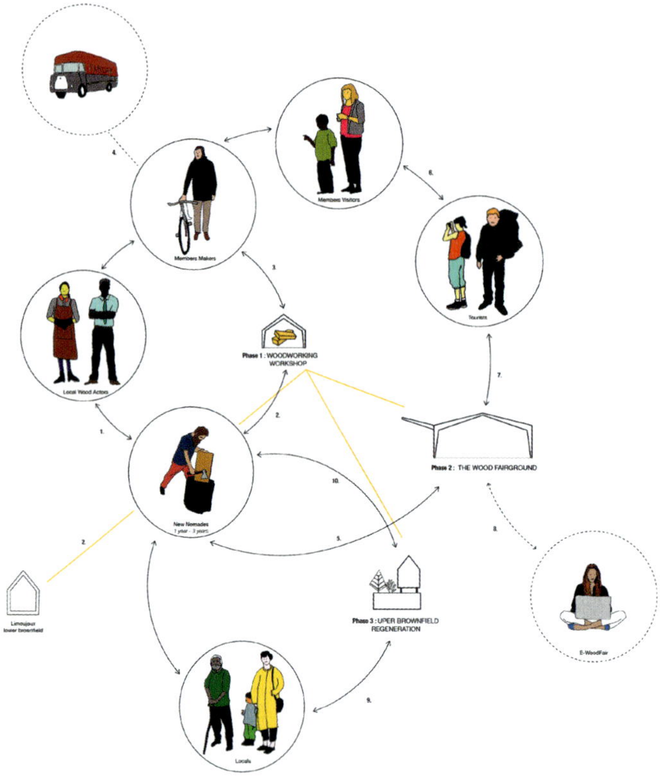

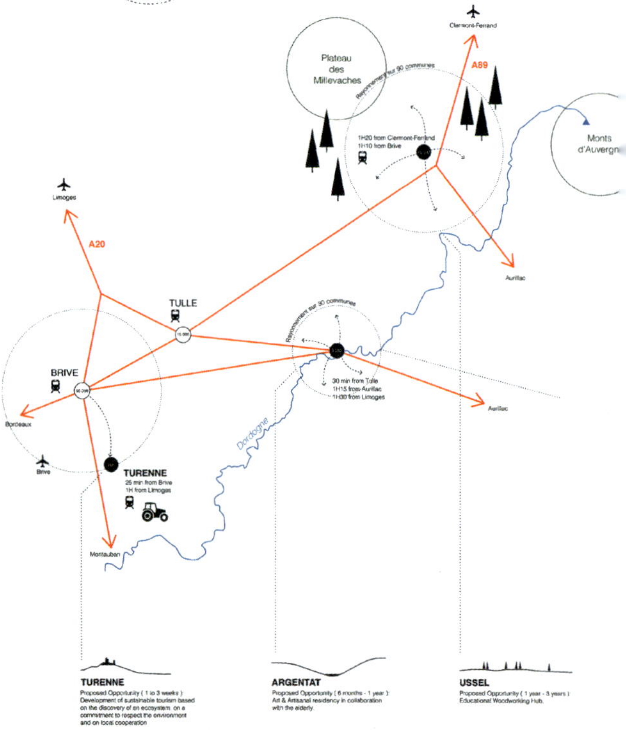

"New nomads", E13 special mention for the La Corrèze site, Pauline Marcombe and Majed Katir.

This project aims to attract an active and optimistic population.

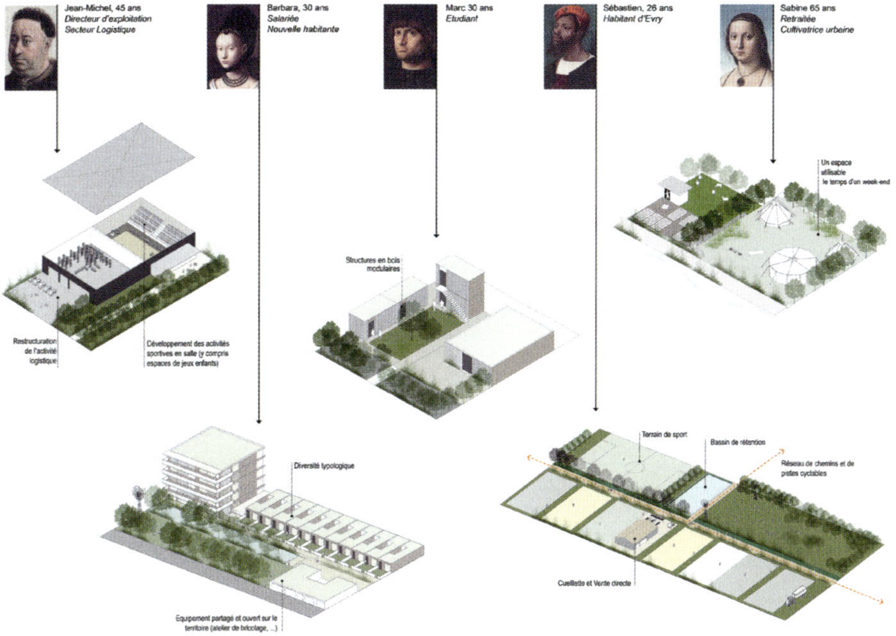

"Terres vives – Les nouveaux communaux " [Terres Vives - The new common lands], E14 runner-up for the Grigny/ Ris-Orangis site, Maia Tüür and Yoann Dupouy.

Based on a progressive approach to property ownership on the site, "Les nouveaux communaux" reestablishes the powerful notion of collective ownership of shared property, affirming the "creative" link between the urban and the boundaries in the making.

POSTFACE

TOMORROW AND THE DAY AFTER

Alain Maugard and Chris Younès

We wrote this article after the completion of the film "En devenir Europan une éternelle jeunesse" [Becoming. Europan, the secret to eternal youth] and the forum held on 14 and 15 December 2018 in the Cité de l'Architecture et du Patrimoine in Paris.

Everything indicates that Tomorrow is already here. The intensity and the speed of the transformations has taken us by surprise: you could say it is accelerating. It is caused by the tyranny of the short-time imposed, in particular, by the digital revolution; it is caused by the scale and speed of the technological progress in the fields of energy, transport, the sharing economy and the circular economy.

In this short temporality, new urban economic models have emerged and quickly taken hold, as have new economic players, both private and public, national and international.

Meanwhile, in contrast, the morphology of cities and territories cannot change rapidly. It has to do with the existing, but needs a long time to transform. This long time as it ensures continuity and stability allows slowness: a hindrance for some, a saving grace for others. This is the power of territories and environments, their biogeographical and cultural heritage.

There is also, and this is crucial, the transitional temporality of the transformation of cities and territories' metabolism. This is when our ways of life evolve, when customs change, it's the time of the Living of the city: it's the time of the rhythm of Nature.

These three temporalities of time exist simultaneously; they interact.

All territories (bus lanes, suburban roundabouts) are not equal in terms of these transformations. Clearly, tomorrow will be easier for some and more difficult for others.

Consequently, it's even more important that we stay the course on our journey towards a day after tomorrow that is harmonious and desirable – a bountiful utopia!

This vision incorporates the environmental, social, economic, political, ethical and aesthetic dimensions of human settlement: the productive city, the evolving (or adaptable) city, the resilient city, the inclusive city, the fair city, the city that protects both individual and collective liberties.

Young students of architecture, urbanism and landscape architecture who have worked throughout this forum have expressed this vision in their own way as we look towards Europan 2050, focusing on the plural dynamics of involvement, sharing and openness.

They have quizzed us about this formidable challenge in which they will be the actors, in order to address new contextualized and translocal scales that are as limited as the Earth itself. They have been called to take part in this debate and they will give it their all: that will be Europan's job and role over the next 15 sessions.

THE AUTHORS

Chris Younès, psychosociologist, Doctor and HDR (Accredited Research Supervisor) of Philosophy, Professor at the École Spéciale d'Architecture, founder and member of the Gerphau think tank (EA-7486) Ensa Paris-La Villette, President of the PhilAU, the Thematic Scientific Network of Philosophy, Architecture and Urbanism (French Ministry of Culture). Co-founder and member of Arena (Architectural Research European Network) and the journal *L'esprit des villes*, member of the Europan Europe Scientific Committee. Most recent published work: *Architectures de l'existence*, Paris, Hermann, 2018.

Alain Maugard, civil engineer, Permanent Secretary of Plan Construction Architecture (Pan: Programme architecture nouvelle [New program for architecture]), Director of Construction (launch of Europan), Director General of the La Défense business center Planning Commission, President of the Scientific and Technical Center for Building (CSTB), President of Europan France, author of several books on sustainable buildings and the sustainable city.

Céline Bodart, architect, graduate of the University of Liege (2010); graduate of the DPEA program specializing in "Architecture and Philosophy" at Ensa Paris-La Villette, and the Masters in Arts and Politics program at Sciences Po Paris; PhD in Architecture jointly awarded by Université de Paris 8 and ULiège in 2018. She currently works as a research assistant in the Faculty of Architecture in ULiège and as an associate lecturer at Ensa Paris-La Villette (Gerphau, EA-7486). Member of the Europan Europe Technical Committee since 2016.

Bernard Reichen graduated from the École Spéciale d'Architecture and co-founded the Reichen et Robert practice with Philippe Robert in 1973. Its transformation of the Leblan spinning mill in Lille and the Grande Halle de la Villette in Paris focused the practice from the outset on the conversion of industrial buildings and territories. Today, this dynamic of "reuse" forms part of a new urban narrative, organized around hypotheses of sustainable development. Bernard Reichen was previously a member of the Europan Europe Scientific Committee. In 2005, he received the Grand Prix de l'Urbanisme, for his work on the future of city-territories.

Didier Rebois, architect, is Secretary General of Europan Europe. He is a lecturer at the École nationale supérieure de Paris-La Villette, and works as an expert and exhibitions commissioner for the Institut de la ville en mouvement [City on the Move].

Xavier Bonnaud, architect, is an associate with the practice Mesostudio. He has a PhD in Urbanism, and is a lecturer at the School of Architecture at Paris-La Villette and the École Polytechnique. He is Director of the Gerphau think tank (EA-7486). Winner of Europan 4 for the Liège site in Belgium, he became a project expert for the technical committees of Europan sessions 9 and 10, and a member of the French jury for Europan 11.

TABLE DES MATIÈRES

Cet ouvrage a été composé en Karmina [Veronika Burian & José Scaglione, 2007] pour le texte
et Futura [Paul Renner, 1927] pour les titres.
Achevé d'imprimer le 25 juillet 2019
sur les presses de l'imprimerie Estimprim à Autechaux
pour le compte des Éditions Parenthèses à Marseille.

Dépôt légal : août 2019.